# THE NATURAL GAS INDUSTRY

# THE NATURAL GAS INDUSTRY: EVOLUTION, STRUCTURE, AND ECONOMICS

Arlon R. Tussing and Connie C. Barlow

BALLINGER PUBLISHING COMPANY
Cambridge, Massachusetts
A Subsidiary of Harper & Row, Publishers, Inc.

iv

International Standard Book Number: 0–88410–975–5

Library of Congress Catalog Card Number: 83–22520

Printed in the United States of America

**Library of Congress Cataloging in Publication Data**

Tussing, Arlon R.
  The natural gas industry.

  Bibliography: p.
  Includes index.
  1. Gas industry—United States—History. I. Barlow,
Connie C. II. Title.
HD9581.U5T87      1983      338.2′7285′0973      83–22520
ISBN 0-88410-975-5

Research was funded in part by the Ford Foundation through the University of Alaska, Institute of Social and Economic Research.

# CONTENTS

### Chapter 6
### A WANING RESOURCE?
### Gas Production in North America:
### Its History and Outlook

### Chapter 7
### GAS DEMAND AND MARKETING
### PRINCIPLES

# LIST OF FIGURES

# LIST OF TABLES

# LIST OF MAPS

# 1 U.S. NATURAL GAS IN PERSPECTIVE

Natural gas—despite its cleaner burning qualities—has always ranked as a poor cousin to oil. It is trickier to transport than liquid or solid fuels, and the costs associated with gas storage prohibit customer stockpiling. Moreover, the industry has yet been unable to secure a niche for gas as a transportation fuel. For all these reasons, refined petroleum products claim the biggest share of energy use nationally and worldwide. (See Figure 1-1.)

Gas, nevertheless, accounts for more than one-fourth of all the primary energy consumed in the United States, and it is the largest domestic source of energy. (See Figure 1-2.) In the 1970s and into the 1980s, gas captured an uncommon amount of political attention as this country suffered through the gas shortages then reeled under a swelling surplus.

Perhaps more than at any other time, the events of the early 1980s demand a skeptical approach to the prevailing wisdom regarding gas-market dynamics and outlook. Recent misjudgments by industry leaders are now painfully evident. Pipeline companies that were sued by angry customers during the gas shortages of the 1970s find themselves in court once again during the early 1980s. This time, it is gas producers who are perturbed as pipelines attempt all sorts of schemes to shed gas-supply commitments that exceed customer demands.

**Figure 1-1.**    U.S. Energy Consumption, 1981
(Including Imports)

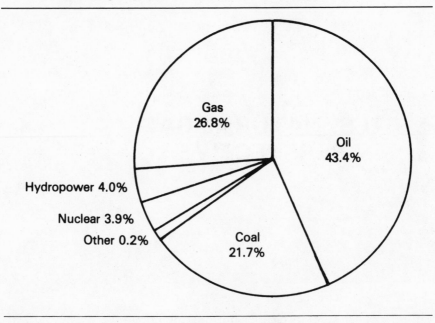

Producers, too, have reason to proceed cautiously. Within a few months in mid-1982, companies drilling for gas free of government price controls (*deep* gas, found below 15,000 feet) watched their anticipated sales at $7.00, $8.00, or even $10.00 per-million-btu succumb to market forces that tolerated at most $3.00 or perhaps $4.00. Even those deep-gas producers whose contracts were negotiated during better times had cause to worry. The most carefully crafted legal covenants were scant shelter from the winds of changed economic circumstances. For nobody can expect to squeeze much money out of a pipeline buyer or a distributor that is financially distressed or, worse yet, bankrupt.

Pipelines and producers are not alone in their distress. Consumers, too, have suffered from swings in gas availability, pricing, and outlook. During the mid-1970s, homes and new industrial plants in most parts of the nation could not obtain gas hookups at any price. But the Natural Gas Policy Act and companion legislation passed in 1978 proved to be more successful than their sponsors had hoped, both in spurring new production and in depressing demand. State-imposed

**Figure 1-2.**   U.S. Energy Production, 1981

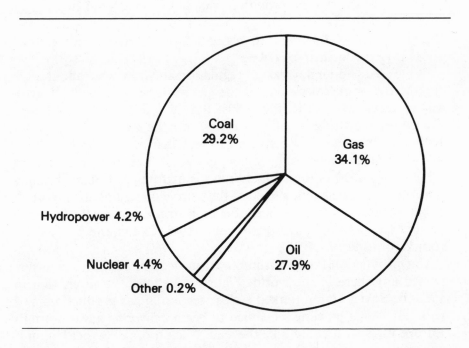

moratoria on consumer hookups fell by the wayside, and industrial consumers curtailed during the 1970s were offered more gas than they wanted; by mid-1982 gas prices had risen to a point where industries found alternative fuels more attractive. Residential customers, too, were hit by painful rate increases soon after they had regained confidence about the gas-supply outlook and were channeling their dollars accordingly.

Meanwhile, federal regulators claimed they were powerless to mitigate the situation, and angry or desperate consumers appealed to Congress. The situation demanded action, but what kind of action, and what were the political obstacles? It had taken several agonizing years and a superhuman lobbying effort before members of Congress from gas-importing and gas-exporting states agreed on a legislative approach in 1978. Five years after the Natural Gas Policy Act became law, opinion was even more divided. No consensus definition of the problem (much less an agreed-upon solution) existed even within the individual sectors of the gas industry. The mesh of federal regulation had grown so complex and had created so many artificial distinctions (intra- versus interstate pipelines, exempt-gas producers versus old-gas producers versus new-gas producers) that suggestions for change ranged from full deregulation of all gas prices and even the deregulation of pipeline rates and service, to regulatory measures far stiffer than those that predated the 1978 act.

What to do about natural gas regulation and how to go about making the day-to-day decisions of the gas business became the biggest energy questions of the mid-1980s. By 1983, natural gas no longer played second-string to oil. It was not only the hottest energy-policy issue, but there was a growing awareness that our nation had direct access to far more natural gas than crude oil. Less than 5 percent of all gas consumed in the United States is of foreign (primarily Canadian) origin, while about a third of the petroleum liquids are imported. What is more, with the exception of a dribble of costly liquefied natural gas (LNG) from Algeria, all of this gas comes entirely from sources outside of OPEC.

Gas production (and consumption) in the United States is unsurpassed anywhere in the world. Figure 1–3 shows, however, that in 1980 the Soviet Union ranked a close second in gas production, and with the Yamal pipeline scheduled to begin delivering gas to western Europe in 1984, it could take the lead. Then too, the Soviet Union is unmatched in proved reserves, while it is widely believed that the major oil-producing countries of the Middle East have huge endowments of as-yet undrilled gas fields. Until those resources become marketable commodities, however, their existence has no real impact on the worldwide flow of energy.

At least for the foreseeable future, therefore, an understanding of the U.S. gas business requires little knowledge of gas resources and institutions outside of our borders. A sufficiently cosmopolitan perspective, indeed, would move the scope of inquiry no further than the limits of the North American continent. To understand the gas business in the United States, one does not, for example, need to know about the resource endowment or royalty rates in Indonesia. The same cannot be said for either oil or coal, which are the major competitors to gas. Figure 1–4 illustrates the dearth of world trade in natural gas compared to its fossil-fuel cousins.

With the exception of certain sections on Canada and Mexico and on world oil prices, this book focuses almost exclusively on the United States. Until and unless the supply situation in North America takes a substantial turn (for the worse or for the better), and until technical

**Figure 1-3.** Worldwide Natural Gas Production and Reserves, 1980.

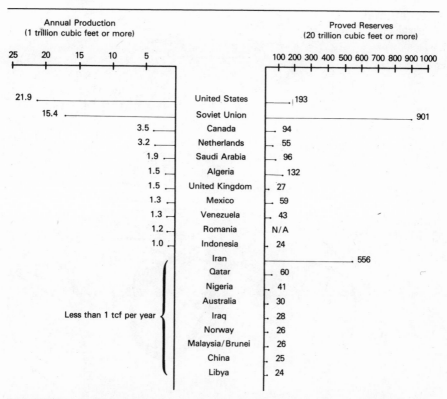

Sources: Production data (including reinjected, flared, and vented gas as well as marketed volumes) is from Energy Information Adminstration, U.S. Department of Energy, *1981 International Energy Annual* (Washington, D.C.: U.S. Government Printing Office, September 1982). Reserves data is from Liscom, William L., ed. (for World Energy Information Services), *The Energy Decade: 1970-1980* (Cambridge, Mass.: Ballinger Publishing Company, 1982).

breakthroughs reduce the costs of gas liquefaction and tanker transport, we suspect that this narrow view will remain both adequate and relevant for U.S. readers.

The text itself is organized both historically and structurally. We were pleased to find that it could proceed chronologically and still provide full and individual treatment of each sector of the industry. For this reason, it is possible to read chapters out of order. We do, however, suggest to those readers seeking an intuitive understanding of the entire gas industry that they make the effort to track the course

**Figure 1-4.** International Trade of Gas, Oil, and Coal, 1980.

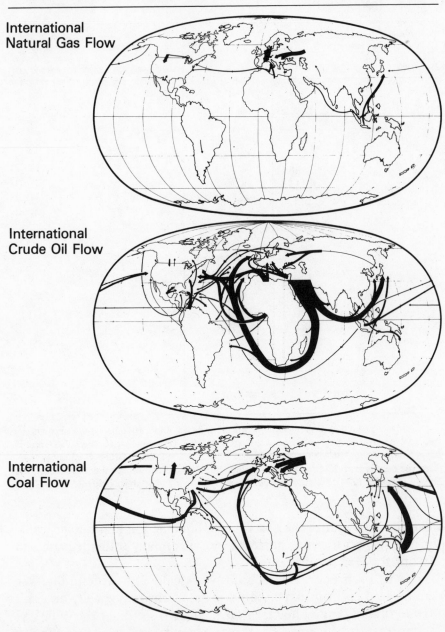

International
Natural Gas Flow

International
Crude Oil Flow

International
Coal Flow

Source: Energy Information Administration, U.S. Department of Energy, *1981 International Energy Annual* (Washington, D.C.: U.S. Government Printing Office, September 1982).

we have marked. Technical terms are italicized wherever they first appear. Although their meanings should be evident in context, the reader may wish to consult the glossary at the end of the text.

There are seven substantive chapters. Five highlight the four main divisions of the gas industry. Chapter 2 explores the roots of gas-distribution companies and the chaos of early attempts by state and municipal governments to regulate their activities. The birth and maturity of long-distance gas-transmission and interstate pipeline companies is detailed in Chapter 3, including anecdotal treatment of each of the companies that now dominate the industry. Chapter 4 examines pipeline-company activities during the 1970s gas shortages, specifically, the plans for acquiring supplemental supplies of gas through LNG imports, coal gasification, and Arctic pipelines—and their quick demise in the 1980s.

Before proceeding into the final two sectors of the industry (gas producers and end-users, Chapters 6 and 7 respectively), the text moves into a purely historical mode. Chapter 5 surveys the growth of government involvement in the industry. One cannot, after all, begin to understand gas production in the United States without a grasp of the regulatory framework that has had such a profound effect on incentives for exploration and development of gas reserves.

The following chapter on gas demand surveys the remaining segment of this vertically disintegrated industry. Our treatment of consumers as coparticipants is a distinct break from the way the industry is usually portrayed. Tradition has divided the gas business into three parts: production, pipeline transportation, and distribution. But all three have lately become aware of the power wielded by the ultimate purchasers of their products and services.

For many years it seemed that consumers would passively accept whatever price resulted from the sum of charges imposed separately by producers, pipelines, and distributors, and that they eagerly would use up whatever volume of gas was sent their way. This was because federal regulators held the price of gas below its market value, and unserved demand therefore always existed at the going rate. But times have changed. Distributors, pipelines, and producers are forced to recognize that their business now depends upon cooperation to keep gas rates competitive with the alternate fuel choices available to consumers.

Chapter 7 is also a short course on gas markets and, more generally, microeconomic principles. The authors intended, in fact, to make this text more than an in-depth review of a single industry. The gas business provides a marvelous case study for understanding fundamental principles of a variety of disciplines. The production chapter, for example, offers a painless introduction to petroleum geology. Chapter 4, through examination of the difficulties encountered by the sponsors of supplemental-gas ventures, covers some essential principles of finance. The remaining chapters provide insights into the arcane world of utility regulation and utility economics.

The final two chapters pull the pieces together. Chapter 8 explores the structural evolution of all segments of the gas industry, with special attention to the forces that fostered or discouraged linkages among the component parts. In Chapter 9, we examine the outlook for the industry and discuss our own views on what is likely and what ought to happen. By this time, the reader should be well-equipped to scrutinize the merits of our forecasts and opinions, based on his or her own political, economic, and even philosophic disposition.

# 2 COAL GASIFICATION IN THE NINETEENTH CENTURY
## Roots of the Gas Distribution Business

### NATURAL GAS AND THE TECHNOLOGY GAP

Natural gas was put to productive use in the United States as early as 1821. Not until the second quarter of the twentieth century, however, did it begin to capture a noticeable share of the energy market. A few people were lucky enough to have a natural seep or "burning spring" in their backyards and the ingenuity to put it to use, but the discovery of natural gas was more frequently an unwelcome event. Coal miners, in particular, dreaded this invisible, explosive vapor.

Even after 1859, when Colonel Drake's first oil well near Titusville, Pennsylvania ushered in the petroleum era, common use of natural gas was half a century away. Gas provided the lift for oil that surged from Drake's well, however, and the Titusville field furnished gas for the nation's first successful natural-gas transportation system. Built in 1872, the wrought-iron pipeline was two inches in diameter and spanned the 5 1/2 miles from well to village.

Gas production and transportation was a risky business—both physically and financially. S.R. Dresser's invention of a leakproof coupling in 1890 was an important breakthrough. Materials and construction techniques, nevertheless, remained so cumbersome that nobody seriously considered bringing gas into any town that was more than a hundred miles from a source of supply.

Added to the technical difficulties of early pipeline construction were the risks that a particular gas resource might disappear soon after the pipe was laid. Indeed, such disappointments were common in the Gas Belt of central Indiana, where shallow reservoirs were tapped and effectively drained within two decades (1886–1907). Not until the 1950s, when wide application was made of modern techniques for estimating reserves (that had gained credibility in the 1940s), did the industry overcome this economic risk.

Louis Stotz and Alexander Jamison, in their classic *History of the Gas Industry* (1938:71), aptly summarize the situation:

> The value of natural gas as a fuel was always recognized, but there was either too much or too little of it at a time; the pressure was variable, there was no way to store it, and with coal and oil so cheap, the economic incentive to overcome the difficulties in handling it was lacking.

During the 1800s, gas was used almost exclusively as a source of light. Electricity, however, captured the lighting market at the turn of the century. Gas companies successfully shifted their promotional efforts to thermal uses and even designed and retailed gas appliances for home use, including water heaters, air conditioners, and cooking ranges. Today, about half of all homes and commercial establishments in the United States use gas for heating or cooking.

Industrial plants and electric utilities, however, account for the biggest chunk of the gas market, consuming more than half of all gas sold in the United States. Glassmaking, paper manufacturing, and petrochemical production, for example, are considered *process-gas* consumers, for which alternative sources of energy are substantially inferior. Other industrial activities, along with the huge electric-utility market, burn gas in order to raise steam for thermal uses or electrical generation. For these *boiler-fuel* markets, gas faces stiff competition with other bulk fuels, primarily residual oil and coal.

Industrial and electric-utility markets for gas were relatively small until after World War II, when *natural gas* became widely available. Before then, the gas industry marketed *manufactured gas*, which by 1850 was already illuminating streets and homes in most large towns east of the Mississippi River. Gas manufactured from coal was practical for these uses, but it made no sense to burn this expensive synthetic fuel in boilers that could burn coal directly.

The prior existence of a strong and widespread manufactured-gas industry made it easy for natural gas to penetrate energy markets in the twentieth century. But until the pipeline boom following World War II, gas produced in association with oil was mostly flared, while *nonassociated gas* was simply left in the ground. Increasingly, U.S. leaders despaired of the irony that gas was manufactured in the city while it was flared in the field.

Throughout the nineteenth century, no clear demarcation existed among the gas industry's production, transmission, and distribution arms. Gas was produced from coal in one or more *gas works* in each urban area. Usually, the same company that manufactured the gas laid the pipeline and distribution system. Today, however, very few of the nation's biggest gas producers invest in long-distance transmission or urban distribution. Companies engaged in the latter two activities, in turn, are rarely affiliated with each other.

The modern gas-producing and -transmission industries are nothing like their predecessors, and the industrywide upheaval of the 1980s may bring massive structural changes once again—with the outcome yet unclear. Technologies have changed, new companies have taken control, and the commodity itself is altogether different. The distribution business, however, harbors strong connections with the past. Distribution companies function today whose corporate charters date from the mid-1800s, and cast-iron mains almost a century old are still carrying gas in many American cities. Then too, distribution companies have largely escaped the reach of national policies and laws. Regulation of this sector has remained almost exclusively the province of state and local governments, which have chosen to act (or not to act) in a variety of ways.

The most important issues now confronting the gas industry are, however, national in scope, and recent debate has focused on the impact of federal policies regarding gas production, transmission, and consumption. For these reasons, this text only touches on the evolution of gas-distribution companies and their state and local regulation. Our emphasis is, rather, on the transformation of the industry's overall structure and the growth of federal regulation of gas production, long-distance transmission, and end-uses.

## SYNTHETIC-GAS MANUFACTURING

The West has known for several centuries that energy-rich vapors can be drawn from liquid and solid hydrocarbons. In 1670, for example, the Reverend John Clayton of Yorkshire recorded his success in distilling flammable vapors from coal. The first instances of gas manufacturing for commercial purposes appeared sporadically in England at the very end of the 1700s.

All organic solids and liquids emit flammable vapors when heated in an oxygen-poor environment. Absent oxygen, the substance will not burn, and its vapors can be collected and used elsewhere as fuel. The flammable vapors comprising manufactured gas are principally *hydrogen* ($H_2$, about 50 percent), *carbon monoxide* (CO, from 5 to 50 percent), and *methane* ($CH_4$, 20 to 30 percent). All of these gases are capable of adding oxygen to their structures and in so doing will emit energy in the form of heat and light.

*Carbon dioxide* ($CO_2$) is also present in manufactured gas. Because it is already fully oxidized (the flammable components of synthetic gas, in fact, turn into carbon dioxide and water when burned), it contributes nothing to the energy value of the gas stream. A high concentration of carbon dioxide is, therefore, undesirable because it dilutes the flammable vapors.

Carbon monoxide, though flammable, has one serious drawback. It is extremely toxic. In the early days of the industry, gas was notorious as a means of suicide and was occasionally the cause of accidental asphyxiation. Even after carbon monoxide-free *natural* gas was introduced, it took a long time for the gas industry to dispel public fears.

Throughout the history of gas manufacturing, a surprising assortment of organic substances has served as feedstock. In addition to coal, gas works used crude oil, refined oil, wood, pine tar, rubbish, and just about any organic substance that was cheap and abundant. A shortage of oil during the Civil War, for example, inspired gas companies in several Confederate states to process cottonseed. In 1849, a plant in Dayton, Ohio, used grease from a nearby slaughterhouse, and the city of Omaha made gas from corncobs in the early 1920s.

Increasing access to cheap and plentiful supplies of natural gas (which is primarily methane) put an end to this creative spirit. The gas shortages of the 1970s, however, revived the market for unconventional feedstocks. Factories and municipalities in the United States and abroad experimented with *methanation* of sewage-plant sludge, feedlot sweepings, and urban garbage.

When manufactured gas began to capture city lighting markets during the first half of the nineteenth century, coal was by far the dominant feedstock, and the distillation technology was primitive. (See Table 2-1 for a chronology of gas use in major U.S. cities.) Around 1870, an abundance of refined petroleum products triggered construction of gas works that used a mixture of coal and oil. One of the lighter distillates of crude oil, *naphtha*, was especially attractive because it had no alternative market and because the resultant *oil gas* contained far less carbon dioxide, carbon monoxide, and sulphur impurities than *coal gas*.

In those days, the only saleable component of crude oil was *kerosene*, which was a popular fuel for home and street lighting. Naphtha, on the other hand, was far too light and explosive to use in lamps and

**Table 2-1.** Introduction of Town Gas in Major U.S. Cities.

| East and Southeast | | Midwest | |
|---|---|---|---|
| Baltimore | 1816 | Cincinnati | 1840 |
| New York | 1823 | St. Louis | 1846 |
| Boston | 1829 | Detroit | 1849 |
| New Orleans | 1832 | Cleveland | 1849 |
| Louisville | 1832 | Chicago | 1850 |
| Pittsburgh | 1836 | Columbus | 1850 |
| Philadelphia | 1836 | Indianapolis | 1852 |
| Providence | 1848 | Milwaukee | 1853 |
| New Haven | 1848 | Toledo | 1854 |
| Buffalo | 1848 | St. Paul | 1857 |
| Rochester | 1848 | Kansas City | 1867 |
| Washington, D.C. | 1848 | Omaha | 1868 |
| Norfolk | 1849 | Minneapolis | 1871 |
| Memphis | 1852 | | |
| Atlanta | 1854 | | |
| Scranton | 1857 | | |

| Pacific Coast | |
|---|---|
| San Francisco | 1854 |
| Portland | 1860 |
| Oakland | 1867 |
| Los Angeles | 1867 |
| Seattle | 1873 |
| Tacoma | 1885 |
| Spokane | 1887 |

was therefore available at bargain rates. Today, however, naphtha is one of the highest-valued products of oil refining; out of it, motor gasolines and jet fuels are drawn. Consequently, gas companies that do utilize naphtha feedstocks during the coldest winter months pay a hefty price for their *peak-shaving* supplies of *substitute natural gas (SNG)*.

In 1854, the United States issued the first patent for the *water-gas* process. Twenty years later, Thaddeus Lowe introduced the more sophisticated *carbureted water-gas* technique. All water-gas technologies injected steam during the later stages of coal or oil distillation. The heat prompted water molecules ($H_2O$) to break apart into component hydrogen and oxygen; the hydrogen atoms then paired, and the oxygen bonded to carbon. This boosted the concentration of free hydrogen and carbon monoxide, thereby enriching the gas vapors.

Coal gas, oil gas, and the various techniques for synthesizing water gas were all still in use during the late 1800s. In selecting a feedstock and technology, a gas company considered the regional availability and comparative costs of coal and oil, the market demand for enriched vapors, and the threat of competing enterprises. Entrepreneurial whim and the inertia of sunk investments powerfully affected the choice of feedstock, as well. For example, by 1890 every gas works in Chicago had abandoned coal gas in favor of water-gas processes. Yet in 1899 eighteen coal-gas plants were still functioning in California, along with ten carbureted water-gas plants and five oil-gas works.

In the last decade of the nineteenth century, the *coke* industry supplied much of the nation's synthetic gas. Flammable vapors (*byproduct coke gas* or *coke-oven gas*) were emitted incidental to the coking process, which transformed wood or coal into almost pure carbon for use in steel manufacturing. By 1930, about 90 percent of all U. S. coke was produced in ovens that jointly manufactured gas, and coke-oven gas accounted for 37 percent of the synthetic gas sold nationwide.

Though natural gas has now totally captured the gas-transmission and -distribution business in the United States, coal gas is still an important fuel in countries (notably Japan) that lack their own resources and must import fuels that are easily transported by sea. During the energy crises of the 1970s, however, many U.S. companies began to experiment with more advanced coal-gasification technologies. The gas shortages and the bleak outlook for domestic supplies fostered an interest in commercial production of a superior grade of coal gas that could be used interchangeably with natural gas. By the time corporate and government interests reached agreement on financing mechanisms, the gas shortages were over. Coal gas has once again been overshadowed by its natural cousin.

## HISTORICAL SHIFTS IN GAS USE AND COMPETITION WITH OTHER FUELS

From its infancy and throughout the nineteenth century, manufactured gas was used exclusively for illumination. Gas works sprung up around the country beginning in the 1820s and soon displaced tallow candles as the chief source of light. Manufactured gas was able to garner a price commensurate with its costs only because its flame was superior to anything else then available, including cheap and abundant whale oil. Whale oil was, in fact, its strongest competitor. Many American cities built their first gas plants in midcentury, when over-hunting was bringing the whaling era to an end.

The reign of manufactured gas as the country's favored illuminant was, however, short-lived. In 1859, the world's first oil well was drilled in Pennsylvania, and by the 1870s, kerosene had captured a large share of the lighting market. Kerosene had undeniable advantages. It posed little danger of asphyxiation, it did not depend upon fixed pipelines, and it burned with a brighter flame.

The Wellsbach mantle lamp introduced in 1885 temporarily restored the primacy of gas as a source of artificial light. This *incandescent* lamp improved the lighting characteristics of gas sevenfold (from 3 to 20 *candles* per cubic foot) because instead of a flame the heat caused a central metal core to glow. By then, however, gas had an even more formidable adversary.

In 1882, New York City fired up the nation's first central electricity generator, making it possible for residents to light their homes with incandescent bulbs. The days of manufactured-gas lighting drew to a close, and the industry had no choice but to nurture other uses for its product. At the turn of the century, gas companies expanded into designing and marketing new appliances. For some companies, in fact, the appliance business became the most profitable part of integrated operations.

Gas irons, gas hair-curling instruments, and other assorted specialties passed in and out of style. It was, rather, the development of appliances for cooking, water heating, and space heating that enabled gas to maintain and increase its share of the nation's energy markets. These appliances owe their existence to Robert Bunsen's 1885 burner design (which premixed gas with air) and to the invention of thermostatic controls, first applied to water heating in 1899.

The transition of gas from an illuminant to a heat source prompted a revolution in measuring standards. While the volumetric standard for gas transactions remained the same, the quality standard denoting energy values changed dramaticaly. Originally, the energy value of gas was expressed in *candlepower*. Conversion to a heat standard came first in Wisconsin, when in 1908 the Wisconsin Public Utilities Commission ordered gas distributors to switch to *British Thermal Units (btus)*. Soon, other states made the change. With the conversion of New York City gas accounts in 1922, the candlepower standard met its ultimate demise. Table 2–2 shows the terms of measurment commonly used in the gas industry today.

The home heating market for gas was limited until manufactured gas was replaced by nontoxic, cheap, and abundant supplies of natural gas. Chicago led the way in 1931. Its gas distributor, Peoples Gas Light and Coke Company, was a cowner in a new thousand-mile

**Table 2–2.**    Natural Gas Measurements.

Cubic foot (cf) is the volumetric unit. Sometimes specified as *standard cubic foot* (scf), it is the amount of gas that would fill a volume of one cubic foot at 60 degrees F. and at the atmospheric pressure at sea level [14.73 pounds per square inch (psi)]. *British thermal unit* (btu) is the heating value unit. One btu is the amount of heat that it takes to raise the temperature of a pound of water by one degree F. (specifically, from 58.5 to 59.5 degrees).

| Volumetric Unit | Heating Value Unit |
|---|---|
| 1 therm | 100,000 btu |
| 1 quad | 1 quadrillion btu (roughly equivalent to 1 trillion cubic feet of gas) |
| 1 cubic foot of natural gas | 1,026 btu |
| 1 barrel of crude petroleum | 5,800,000 btu |
| 1 pound of bituminous coal | 12,000 btu |

| Volumetric Unit of Oil or Coal | Volumetric Equivalent in Cubic Feet of Gas |
|---|---|
| 1 barrel of crude oil | 5,650 cubic feet of gas |
| 1 ton of bituminous coal | 23,400 cubic feet of gas |

pipeline that connected the Windy City to the world's then-largest gas producing area—the Panhandle and Hugoton fields of south-western Kansas and adjacent sections of the Texas and Oklahoma panhandles.

A brief setback in consumer demand for gas followed a 1937 explosion of a gas furnace in a Texas school. The mishap killed almost 300 children and adults. By the end of World War II, however, the fright had subsided, and the gas industry began to flourish. For the next thirty years, homeowners and industries in the United States had access to an abundant and inexpensive fuel. End-use prices in the range of 10 to 30 cents per thousand cubic feet (mcf), and for a quality of gas that was twice as potent as its synthetic predecessor, meant that postwar consumers paid only a tenth or a hundredth of what previous generations had to pay.

## THE ROOTS OF GAS-INDUSTRY REGULATION

Government intervention in the affairs of the gas industry reaches back to the earliest years of gas use. Only the producing sector, however, has had a consistently negative attitude about the governmental role. Gas distributors in the 1800s (and, later, interstate pipelines) actively sought government assistance and protection.

Synthetic-gas companies petitioned municipalities and state legislatures for *eminent domain* authority to cross private property and to lay pipelines under city streets and rights-of-way. Because of the scale and fixed nature of gas-plant investments, these companies also wished protection from excessive competition. They requested *franchises* from state and local authorities by which the government agreed to refrain from granting similar rights and privileges to potential competitors. The need for eminent domain authority and the notion that gas distribution is a *natural monopoly* still powerfully influence the structure and behavior of the industry.

Two other factors prompted governmental participation in gas affairs. First, municipal governments were important consumers of gas. A contract to furnish street lighting often was crucial to the economic viability of the system. Because street lighting was the first large-scale use of gas, some municipal governments chose to enter the gas manufacturing and distribution business themselves—and some are still participants today.

Second, the sums of money required for investment in gasification and distribution equipment compelled prospective owners to incorporate. But only a few states enacted general incorporation laws before the late 1800s. Instead, legislatures conferred corporate status on a case-by-case basis through issuance of special *charters*. Many of the nation's distribution companies still operate today under charter status rooted in the last century.

Legislatures were hesitant to unleash the powers of incorporation because the nation as a whole was suspicious of this new form of business organization. Corporate status encouraged passive ownership, insulated stockholders from liabilities borne by the corporation, and provided an opportunity for the company to amass substantial wealth. When states granted privileges of incorporation to gas companies, and especially when governing bodies issued monopolistic franchises and authorized access to public corridors, they therefore instituted safeguards to protect the public interest.

Most of the special terms inserted in charters, contracts, statutes, and ordinances were intended to thwart the inherent evils of monopolies that, unfortunately, seemed to go hand-in-hand with economies of scale and other benefits that argued in favor of the franchise. Through the years, a distinct body of law and economic theory has evolved regarding the proper role of government in regulating gas, electricity, and other franchised enterprises "affected with a public interest." *Public utility* regulation encompasses three arenas: (1) preventing price gouging through *rate regulation*, (2) ensuring acceptable safety standards and quality control of services provided, and (3) obligating monopoly companies to provide service to all interested parties without discrimination.

Of the three, rate regulation has captured the greatest attention and has given rise to the largest body of law and theory. The cornerstone of utility rate regulation is the *cost-of-service* principle. Legislated monopolies that distribute a vital commodity, such as gas, are allowed to charge rates that reflect only the actual cost of providing the service, including a "just and reasonable" return on investment sufficient to sustain stockholder and lender interest in the venture. Determining the actual cost, however, and translating that cost into billings for different classes of customers, turns out to be a complicated and controversial exercise.

Prior to passage of the 1938 Natural Gas Act, there was no federal presence in the gas industry. Government surveillance was strictly at the local and state level. Accordingly, government and industry roles and powers varied from one state to the next (and even from city to city), limited only by occasional judicial admonitions to keep government actions within the bounds of the U.S. Constitution.

Perhaps the best way to understand the early roots of utility regulation, as well as the evolution of the gas distribution business, is to take a detailed look at the history of the gas industry in Chicago. The following chronology is drawn from a 1925 publication of Peoples Gas Light and Coke Company by Wallace Rice entitled *Seventy-Five Years of Gas Distribution in Chicago*.

## THE CHICAGO CASE

Chicago's first gas works and distribution company received a charter from the Illinois legislature in 1849. The Chicago Company secured the right to operate "with perpetual succession." As one of America's first limited-liability corporations, a cautious legislature set a $300,000 ceiling on the amount of stock that could be issued.

The charter established only a limited franchise; the legislature assigned the company an exclusive right to supply Chicago with gas for only ten years. Despite the ten-year guarantee of monopoly status, the charter did not include any rate stipulations. The state action also granted the company sweeping rights to lay gas mains and pipes anywhere under Chicago's streets without specific city consent, so long as no permanent damage ensued.

The municipality, however, was not powerless. Construction financing depended upon the company's securing a contract to provide street lighting. This it received, but the city extracted a firm price of $2.50 per mcf (although it did agree to waive property taxes for the first five years of operation). Still, no rate protection was afforded private customers, and the company fired up its plant in 1850, charging private parties $3.00 per mcf.

In 1855, the Chicago Company secured a charter amendment that raised the ceiling for capital stock and thereby enabled the company to expand its distribution network. At the same time, the legislature revoked its commitment to refrain from chartering competitors, and it awarded Peoples Gas Light and Coke Company a similar charter with a stock ceiling of $500,000. This time, the legislature inserted language that put a ceiling on company rates. Peoples was allowed to charge no more than $2.00 per mcf for servicing city lights and no more than $2.50 private consumption. Neither the abrogation of Chicago Company's monopoly position nor the legislated price competition in fact materialized. Money had become extremely tight, and it was not until 1861 that the new company secured financing.

The entry of Peoples into the Chicago gas business was followed over the next twenty years by that of three other companies organized under a general incorporation statute enacted in 1872. The newcomers posed no threat to the established companies, however, because their franchises restricted activities to the suburbs and outlying towns.

Meanwhile, the Chicago Company and Peoples had adopted a "live-and-let-live" approach. Both scrupulously avoided competition by carving out separate territories. Peace was shattered in 1882 when the Chicago Assembly granted Consumers Gas Fuel and Light Company a franchise and the new company chose to enter Peoples' service area. Intense and debilitating competition racked Chicago's gas industry for the next fifteen years.

By the time Consumers got into the gas business, the state had disclaimed any power to interfere in ratemaking. An 1865 statute removed the stock limitations imposed in Peoples' charter and expressly delegated to the city all powers to regulate rates, save one; state law established a minimum rate of $3.00 per mcf. By the time rate wars began in earnest, even this floor had fallen through, and rate jurisdiction was strictly within the province of the Chicago Assembly.

In the ordinance that gave birth to Consumers, the municipality expressly chose to foster competition. In addition to granting the new company the right to invade markets held by Peoples and the Chicago Company, it prohibited Consumers from transferring any of its rights, joining with any other company to fix rates, or merging with competitors. It thus ensured that the various companies would not create a citywide monopoly on their own.

Because the ordinance set a ceiling for Consumers' services that was 25 cents below the going price, a rate war was inevitable when the company set up operations in Peoples' traditional service area. The fight, however, was short-lived. Competition proved too harsh for the fledgling company; Consumers went bankrupt and its creditors confined company activities to gas manufacturing, selling the output to Peoples and Chicago Company.

Rate wars really got underway in 1886, when the Chicago Company invaded Peoples' territory, thus pitting two established businesses against one another. During the next ten years, Chicago and Peoples (along with several smaller firms) made repeated attempts to merge, but the courts held up or scuttled the plans. The rate wars peaked when the city assembly, once again, stimulated competition by awarding a new franchise. Ogden Gas Company received a fifty-year franchise in 1895, with the usual prohibition against merger. The franchise document set a ceiling price of 90 cents per mcf, which was 10 cents below the prevailing rate. The city involved itself even more deeply in company affairs by inserting a provision for a 3.5 percent tax on Ogden's gross income.

Following Ogden's entry, Chicago rates dipped as low as 40 cents per mcf—substantially less than the $3.00 that prevailed when gas was first sold or the $4.00 price induced by Civil War shortages and war taxes. While some of the price reduction was attributable to economies of scale, improved technologies, and amortization of gas plant, there was no question that companies selling gas at 40 cents were doing so at a loss.

Two years after Ogden received a franchise, the Illinois legislature intervened, passing a law that expressly permitted any gas company to sell its properties and franchises to another gas company in the same city. The major companies in Chicago acted immediately, merging under the name of Peoples Gas Light and Coke. Political realities prompted inclusion in the merger document of a company commitment to sell its product at the lowest rate then current ($1.00 per mcf).

Thereafter, it was customary for the company to confer with city government when changing its rates. Custom was indeed important; not until 1905 did the Illinois Legislature again grant municipalities the power to oversee gas rates. That action, however, was ruled invalid by the courts, so in 1913 Illinois established a state public utility commission vested with ratemaking authority.

## STATE REGULATION OF PUBLIC UTILITIES

The nation's first public utility commissions and public service commissions, (PUCs and PSCs) were born in 1907 in New York and Wisconsin. These two bodies, along with the California and New Mexico commissions and those in several other states, have emerged as the nation's leaders with respect to involvement in gas industry questions

**Table 2-3.**   State Public Utility Commissions and
Their Jurisdictions.

| State and Commission | | Regulation of: Gas Distribution | Gas Transmission | Jurisdiction over Publicly-Owned Utilities |
|---|---|---|---|---|
| Alabama | Public Service Commission | Yes | No | No |
| Alaska | Public Service Commission | Yes | Yes | No |
| Arizona | Corporation Commission | Yes | Limited | No |
| Arkansas | Public Service Commission | Yes | Yes | No |
| | Commerce Commission | No | No | Yes |
| California | Public Utilities Commission | Yes | Yes | No |
| Colorado | Public Utilities Commission | Yes | Yes | a |
| Connecticut | Public Utilities Commission | Yes | Yes | Limited |
| Delaware | Public Service Commission | Yes | No | b |
| District of Columbia | Public Service Commission | Yes | N.A. | N.A. |
| Florida | Public Service Commission | Yes | Safety | No |
| Georgia | Public Service Commission | Yes | No | No |
| Hawaii | Public Utilities Commission | Yes | No | No |
| Idaho | Public Utilities Commission | Yes | Limited | No |
| Illinois | Commerce Commission | Yes | Yes | No |
| Indiana | Public Service Commission | Yes | Limited | Limited |
| Iowa | Commerce Commission | Yes | Yes | No |
| Kansas | Corporation Commission | Yes | No | No |
| Kentucky | Public Service Commission | Yes | Yes | Limited |
| Louisiana | Public Service Commission | Yes | Yes | No |
| Maine | Public Utilities Commission | Yes | Yes | Yes |
| Maryland | Public Service Commission | Yes | Limited | c |
| Massachusetts | Department of Public Utilities | Yes | Yes | Limited |
| Michigan | Public Service Commission | Yes | Yes | No |
| Minnesota | Railroad and Warehouse Commission | No | No | Yes |
| Mississippi | Public Service Commission | Yes | No | Limited |
| Missouri | Public Service Commission | Yes | Yes | No |
| Montana | Board of Railroad Commissioners | Yes | No | Limited |
| Nebraska | Railway Commission | No | Limited | d |
| Nevada | Public Service Commission | Yes | Yes | No |
| New Hampshire | Public Utilities Commission | Yes | Yes | Limited |
| New Jersey | Department of Public Utilities | Yes | Limited | Limited |
| New Mexico | Public Service Commission | Yes | No | No |
| | Corporation Commission | No | Limited | Yes |
| New York | Public Service Commission | Yes | Limited | Yes[e] |
| North Carolina | Utilities Commission | Yes | Yes | No |
| North Dakota | Public Service Commission | Yes | Yes | No |
| Ohio | Public Utilities Commission | Yes[f] | Yes | No |
| Oklahoma | Corporation Commission | Yes | Limited | No |
| Oregon | Public Utility Commission | Yes | No | No |
| Pennsylvania | Public Utility Commission | Yes | No | Limited |
| Rhode Island | Division of Public Utilities | Yes | Yes | No |

**Table 2-3.** *(continued).*

| State and Commission | | Regulation of: Gas Distribution | Gas Transmission | Jurisdiction over Publicly-Owned Utilities |
|---|---|---|---|---|
| South Carolina | Public Service Commission | Yes | Yes | Tel. |
| South Dakota | Public Utilities Commission | No | No | Yes |
| Tennessee | Public Service Commission | Yes | Yes | No |
| Texas | Public Utilities Commission | No | No | Limited |
| Texas | Railroad Commission | Limited | Yes | No |
| Utah | Public Service Commission | Yes | Yes | No |
| Vermont | Public Service Board | Yes | Yes | Yes |
| Virginia | Corporation Commission | Yes | Yes | No |
| Washington | Utilities and Transportation Commission | Yes | Yes | No |
| West Virginia | Public Service Commission | Yes | Limited | Yes |
| Wisconsin | Public Service Commission | Yes | Limited | Yes |
| Wyoming | Public Service Commission | Yes | Yes | Yes[g] |

Source: American Gas Association, *Gas Rate Fundamentals*, third edition (Arlington, Va.: American Gas Association, 1978).

[a]Municipally-owned electric and gas utilties outside corporate limits.

[b]Electric co-operatives only.

[c]Electric and gas utilities only.

[d]Limited Jurisdiction over electric transmission construction for safety.

[e]Except water utilities and municipally-operated local transit.

[f]Municipalities have original jurisdiction over rates. Appellate jurisdiction in Commission. Original jurisdiction in Commission when municipality does not exercise authority to set rates.

[g]Except utilities owned and operated by Municipalities.

of national scope. New York in particular is a frequent *intervenor* in matters before federal administrative and judicial bodies.

Commissions with jurisdiction over gas distribution now exist in forty-eight states. (See Table 2-3.) Some states, however, have vested this power in "corporation commissions," "commerce commissions," or even "railroad commissions"—most notably, the Texas Railroad Commission. Most state commissions also regulate pipeline companies whose activities are confined to a single state. *Intrastate* gas-transmission companies transport gas produced within the state or received at the border to *city-gate* connections with distributors. The states vary widely in the degree to which rate-making powers are delegated to municipalities. Texas and Ohio, for example, delegate substantial rate-making authority to local governments.

Many states assert no jurisdiction over municipally-owned facilities. In 1834, the City of Philadelphia built the first municipally owned gas works. Today, about a third of the 1600 companies distributing gas in the United States are government-owned.

The reach of a state's powers ends at its borders. The U.S. Constitution bars state and local governments from regulating business transactions that cross state lines, whether or not the federal government has chosen to assert jurisdiction. Federal regulation of interstate gas transmission did not begin until 1938, when Congress passed the Natural Gas Act.

During the early years of the gas industry, there was no compelling reason for federal oversight because it was physically impossible to ship gas more than fifty or a hundred miles. Technical innovations, however, sparked the emergence of the long-distance gas transmission industry in the second decade of the twentieth century. As interstate commerce boomed, state regulatory commissions (Constitutionally barred from controlling upstream rates) pressed for federal oversight. In 1938, Congress authorized the Federal Power Commission to regulate interstate gas transmission.

The next chapter explores the technological and other events that gave birth to the interstate transmission arm of the gas industry. The text also outlines the stages of pipeline growth in the United States which, in turn, sets the foundation for investigating the genesis and evolution of federal regulation.


## REFERENCES

Rice, Wallace. "75 Years of Gas Service in Chicago." Chicago: The Peoples Gas Light and Coke Company, 1925.

Stotz, Louis, and Alexander Jamison. *History of the Gas Industry*. New York: Press of Stettiner Brothers, 1938.

# 3 THE TRANSITION FROM MANUFACTURED TO NATURAL GAS
## Evolution of the Gas Transmission Industry

### THE TECHNOLOGICAL CHALLENGE

For many years, oil producers flared natural gas in the field or sold it to *carbon-black* plants for mere pennies per thousand cubic feet, while urban dwellers typically bought manufactured gas at prices of a dollar or more. Willing sellers and willing buyers remained apart until technical advances made it possible to transport natural gas safely and economically.

Until recently, gaseous fuels were far more difficult and costly to move than were oil and coal. Liquid and solid commodities can be poured or dumped into barrels, bins, or tanks and then carried to market by highway, rail, or sea; but the only efficient way to move gas is by fixed, high-pressure pipelines.

Pipeline transportation is not necessarily a disadvantage. Today, for example, pipelines are the preferred mode for overland shipments of oil, while *slurry* pipelines are potentially the cheapest way to move large volumes of coal. But it was not until the second quarter of the twentieth century that pipelines became a practical means for moving natural gas.

Before there was a way of getting gas to market, there was little point in looking for it. Natural gas was discovered, nevertheless. It frequently turned up, unsought and unwanted, in water wells, coal mines, and even cellars. More important, natural gas was a universal but generally unwelcome byproduct of the search for oil.

If a driller struck *nonassociated gas* instead of oil, the find was not a whole lot more valuable than a dry hole. On the other hand, *casing-head gas*, which is almost always associated with oil production, created dangerous operating conditions around drilling and production ·igs, and occasional well blow-outs and fires. It took several decades before oil industry pioneers realized that this *dissolved and associated gas* was, in fact, an indispensible aid to their operations. Expansion of the gas in the reservoir and well bore provided most of the energy that brought oil to the surface in the first place. Unless the gas was conserved (and reinjected), oil recovery ultimately suffered.

Although J.P. Lesley postulated in 1865 (only six years after oil was first drilled commercially) that maintenance of gas pressure within a reservoir was essential for sustained oil production, it took more than half a century for that insight to shape oil-field practices. Some oil producers in Ohio instituted gas reinjection as a method of *secondary oil recovery* in 1903, but it was not until the 1930s and 1940s that this approach became widespread.

In 1913, L.G. Huntley of the U.S. Bureau of Mines published a paper that rediscovered "the gas concept." Thirteen years later, the Federal Oil Conservation Board (which functioned between 1924 and 1930) issued a report concluding that indiscriminate flaring of associated gas reduced ultimate oil recovery. A supplemental document followed that measured the relationship between gas flaring and oil recovery based on actual field experience. C.E. Beecher and I.P. Parkhurst coauthored the study, under the direction of Henry L. Doherty, the "father of petroleum conservation." A 1929 report by H.C. Miller of the U.S. Bureau of Mines ("Function of Natural Gas in the Production of Oil") laid any remaining doubts to rest; proponents of gas conservation had no credible opposition.

Burgeoning reserves of both associated and nonassociated gas, and the enormous waste of energy occasioned by gas flaring, spurred the search for improved transmission technologies. State government restrictions on gas flaring sharpened these incentives. The first pipeline that connected San Francisco with reserves 300 miles away was completed in 1929, just before California's Gas Conservation Act went into effect.

A further impetus to development of long-distance pipelines was the *common-law* principle pertaining to oil and gas ownership. Methods for measuring the amount of petroleum in place, or recoverable reserves, even today are difficult and lack precision. Several decades ago, a crystal ball worked just about as well as scientific calculus. This problem, added to the fact that petroleum resources are mobile and can migrate underground across property lines, led the courts in the late 1800s to fall back on English law pertaining to wild game. Oil and gas ownership was (and still is) determined by the *Rule of Capture*. Quite simply, oil and gas belongs to whoever produces it.

The Rule of Capture gave rise to some decidedly poor field practices. Where a gas reservoir underlay several separately owned properties, as was common in the Panhandle region of Texas, Oklahoma, and Kansas, each operator had a powerful motivation to commence production as soon as possible and to punch as many holes as company finances allowed. An owner was almost forced to get into business as soon as his neighbor started drawing on the reserves. The frenzied competition of producers to lift as much oil as possible before their neighbors drew it left little time to develop economic uses for the associated gas.

State governments were not, however, powerless to contain the chaos created by the Rule of Capture. Beginning in the 1910s with legislation in Oklahoma and Texas, state governments asserted jurisdiction over various aspects of petroleum production in order to prevent physical, and later economic waste. Conservation statutes gave regulatory boards the power to prescribe well spacing and density and to institute pro-rata cutbacks in production if a surplus threatened to unleash cutthroat competition. This practice of *market demand pro-rationing* not only provided market stability but it protected the *correlative rights* of multiple owners drawing from a common pool.

At first geared strictly to the problems of crude-oil production, states later instituted laws limiting and prohibiting gas flaring. Some states (though notably not Texas) instituted compulsory *unitization*, and most enacted *common purchaser* legislation. Nevertheless, it took a long time for some of these regulatory initiatives to bring order into the business of oil and gas production. Moreover, when the early 1980s gas glut struck an industry that had grown complacent about the marketability of its product, state regulators found that existing laws and programs were inadequate.

The Rule of Capture was not the only force for technological advance in pipeline manufacturing during the early days of the gas industry. Consumers, too, were eager to be hooked up to this cheaper, cleaner, and safer fuel. By the turn of the century, manufactured gas had become an important source of energy in most U.S. cities. With the distribution infrastructure already in place, a strong market for natural gas was assured. A prospective gas pipeline, therefore, faced few of the market-entry risks common to new businesses.

Natural gas offered some distinct advantages over the manufactured variety. It contained no poisonous carbon monoxide, and it did not blight the atmosphere with the soot and sulphur compounds that spewed out of gas works. Natural gas also offered twice as much energy per unit volume (about 1000 btu/mcf) as its synthetic cousin.

Switching to richer gas, however, was not without problems. Distribution companies had to adjust the orifices and regulating devices on existing appliances. Nevertheless, natural gas offered so many advantages that these logistical obstacles were nowhere a lasting deterrent. Utilities deployed special crews to enter homes and convert appliances, and fashioned detailed strategies for effecting the change neighborhood by neighborhood.

Once natural gas began to flow from remote regions, it took several decades for the new energy source to capture a chunk of the markets traditionally held by oil and coal. Many distributors sold a mixture of synthetic and natural gas, and only later did they fully convert to natural sources. Detroit, for example, got its first dose of natural gas in 1848, but ninety years rolled by before the distributor had completely eliminated manufactured gas from the sales-gas mixture.

Even after the transition was complete, many distributors still used their old gas works for *peak-shaving* during the winter months of greatest demand. This seasonal increment of synthetic gas, blended into natural stocks, bolstered company supplies without seriously degrading gas quality. Today, underground storage, liquefied storage, and *propane-air* plants have captured most of the peak-shaving market, but ancient gas works are still maintained on some systems for emergencies.

By 1930, natural gas accounted for about four-fifths of all distributor sales in the United States. In many cities of the Southwest and Plains states, synthetic gas systems had never developed, and natural gas was sold undiluted from the beginning. The transition was, however, prolonged in the Northeast, where distributors had pioneered the synthetic gas industry. The very last to convert were cities in the coal-producing region of Appalachia. In 1930, forty percent of the gas consumed in Pennsylvania was still manufactured from coal.

A major reason for the slow death of the synthetic-gas industry was the long time that it took for gas-transmission companies to build enough pipelines to satisfy latent demand. Improvements in pipeline technologies also came in fits and starts. The first gas pipeline to span a distance of more than a hundred miles was built in 1891. It carried gas 120 miles to Chicago from fields in central Indiana, and it did so without compressors. The gas happened to come out of the ground at a high enough pressure—525 pounds per square inch (psi)—to ensure an acceptable rate of flow over the entire distance. The amount of gas that a pipeline can deliver is a function of its diameter and the horse-power and spacing of its compressor stations. Today, most of the major trunklines in the United States operate at pressures in the range of 1,000 psi and receive pressure boosts at least every 100 or 200 miles.

Primitive gas compressors were functioning as early as 1880. Compressor technology, however, was not the limiting factor to industry growth. Pipelining techniques were a far greater stumbling block in the early days of the gas transmission business. Progress depended upon strengthening pipe seams, couplings, and the steel itself in order to withstand the high pressures so essential to project feasibility.

In addition, safety is far more difficult to achieve for gas than for oil pipelines. A small hole or crack in an oil pipeline may be troublesome to detect, clean up, and repair, but it is far less hazardous than a hole in a gas pipeline. Especially in high-pressure pipelines, any injury poses the threat of explosion, and the force may induce a rupture that, in a flash, can propagate itself upstream for a mile or more.

Cast-iron pipe has been available from the beginning of the gas industry, replaced by steel in the last decade of the nineteenth century. Even then, pipe strength was substantially limited because it was rolled from flat sections; stronger or thicker steel did not improve a pipe's ability to withstand high pressures so long as the seam was much weaker. Seam-sealing and pipe-joining techniques moved forward in 1911 with the introduction of oxyacetylene welding, followed by electric welding in 1922. Finally, pressure gas welding employed during and after World War II, combined with manufacturing breakthroughs that cut the costs of steel procurement (by providing greater strength to thinner walls), paved the way for a long-distance pipeline construction boom.

**Table 3-1.**   Growth of Pipeline Capacities.

| Year (circa) | Pipeline Diameter | Design Pressure[b] |
|---|---|---|
| 1930 | Up to 20 inches | 500 psig |
| 1948 | Up to 26 inches | 800 psig |
| 1960 | Up to 36 inches | 1000 psig |
| 1975 | Up to 42 inches | 1260 psig |
| 1980[a] | Up to 56 inches | 2000 psig |

[a]These figures apply to frontier projects for shipment of gas from the Canadian and Alaskan Arctic, none of which materialized by the early 1980s. New construction of conventional pipelines is still measured in the 42-inch range.

[b]Pressure is measured in *pounds per square inch* (psi). Objects at sea level are subjected to an *atmospheric pressure* of about 14.7 psi (which results from the weight of several miles of air resting on the earth's surface). Instruments designed to measure artificially induced pressures like those inside gas pipelines, record or *gauge* pressures in excess of this ever-present atmospheric pressure (psi*g*). *Absolute* pressure measurements include the 14.7 psi exerted by the atmosphere (psi*a*). Hence, 1680 psig is the same as 1694.7 psia.

Together, these advances in pipe rolling, metallurgy, and welding swept aside the physical barriers that separated eager customers from producers bloated with gas. As technologies continued to improve, the industry utilized ever-increasing pipe diameters and pressure standards, which in turn bolstered system capacities and economies of scale. Table 3-1 displays the course of these advances in pipeline capacities.

## GAS TRANSMISSION BEFORE 1925

Significant urban use of natural gas began in three regions. Casinghead gas from the Buena Vista field of the **San Joaquin Valley** flowed to Los Angeles beginning in 1909. Gas produced in the oil fields of **eastern Kansas and Oklahoma** found limited markets in small towns within a hundred miles. The biggest producing area, however, was the **Appalachian** region, where the nation's first pipeline had, in fact, been built. Oil and gas fields were scattered throughout the Appalachian states, roughly within a triangle bounded by Buffalo, Columbus, and the Virginia/West Virginia border. (Map 3-1.) Pittsburgh and Cleveland supported the biggest gas markets.

**Map 3-1.   Long-Distance Gas Transmission before 1925.**

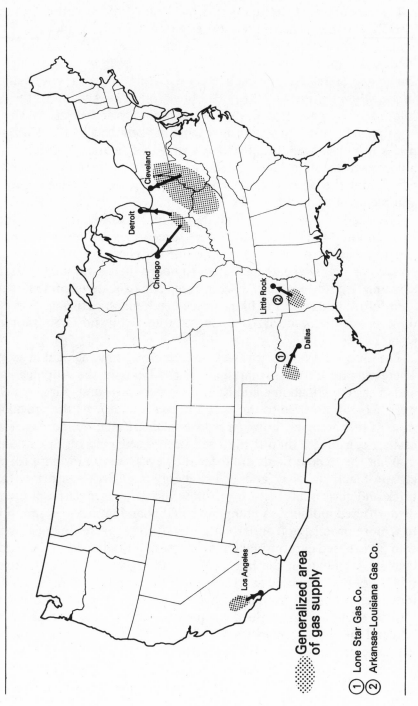

Generalized area
of gas supply

① Lone Star Gas Co.
② Arkansas-Louisiana Gas Co.

The Standard Oil Company of New Jersey and Columbia Gas and Electric Company distributed more than half of all Appalachian gas sold. The latter firm (now **Columbia Gas System**) is still a major force in the gas transmission and distribution business in the eastern half of the United States. Standard Oil's involvement, however, ended in the 1930s, when the Public Utility Holding Company Act forced the dissolution of interstate corporate empires in the gas and electric business.

Appalachian gas production reached its zenith in 1917. Although prices doubled over the next five years, this financial stimulus was not enough to offset the production decline. Indeed, there was little doubt that the doomsayers were correct and that the age of cheap and plentiful natural gas was just about over.

The steady rise in gas demand, despite higher prices, made the disaster even more imminent. Cities were forced to reintroduce manufactured gas as an additive to stretch out supplies. West Virginia and Oklahoma made repeated attempts to hoard their dwindling reserves, enacting legislation that discouraged out-of-state shipments and "wasteful" consumption like carbon-black manufacturing. Some of these efforts were declared unconstitutional by the U.S. Supreme Court.

The bleak outlook for gas was, nevertheless, only regional in scope. There was no question that shallow gas fields in the Appalachians and in eastern Oklahoma and Kansas were on the wane, but new fields were being discovered to the west and to the south. At the prevailing state of technologies, however, those fields offered no more relief to eastern consumers than if they had been discovered on the moon.

What the remote fields did offer, however, was a new base for the carbon-black industry and other energy-intensive enterprises (like brick and glass manufacturing) whose relatively meager plant investment offered mobility. A company could abandon facilities and move to a more promising frontier without suffering severe financial hardship. Gas-based carbon-black manufacturing originated in the Appalachian gas fields in the late 1800s. Although carbon-black (essentially soot that was used as a colorant in paints and printing inks) had long been produced from other hydrocarbons, low-temperature combustion of natural gas yielded a superior product that was high in carbon and low in ash impurities. Natural gas, therefore, rapidly replaced kerosene as the dominant feedstock of the black industry.

During the 1920s, the demand for carbon-black expanded beyond traditional end-users. Carbon-black moved into the rubber industry where it became a vital reinforcing agent. A shortage of carbon-black during World War II, for example, meant that civilians had to accept noticeably inferior grades of automobile tires and rubber boots.

As a "scavenger" industry, successful only where higher-value users could not compete for feedstocks, carbon-black manufacturers repeatedly moved their operations to the most remote gas fields. By the 1950s, however, virtually no producing area in the United States was untouched by the interstate gas transmission network. The carbon-black industry switched back to oil and increasingly moved overseas. By 1981, this industry accounted for only about .1 percent of domestic gas consumption.

## THE PIPELINE BOOM OF THE LATE 1920s

Long-distance gas transmission finally became practical in the late 1920s with advances in pipe technology. Between 1927 and 1931, about a dozen major transmission systems emerged, each with pipe diameters of about 20 inches and covering distances in excess of 200 miles. These systems tapped gas in three hydrocarbon provinces: the Panhandle-Hugoton fields, the Monroe field in Louisiana, and California's San Joaquin Valley. Table 3–2 and Map 3–2 provide details of these developments.

The **Panhandle-Hugoton** fields represent the biggest single accumulation of gas ever developed in the Western Hemisphere. Stretching almost 300 miles from the northern portion of the Texas Panhandle, across Oklahoma, and into southwestern Kansas, these midcontinent fields are composed mostly of nonassociated gas. Key discoveries in the Panhandle field occurred between 1910 and 1920, with the even bigger Hugoton field discovered in the decade that followed. Table 3–3 shows that the original *recoverable reserves-in-place* were probably about 117 tcf—enough to supply the whole nation for six years at the rate of consumption that prevailed in the 1970s and into the 1980s.

The **Cities Service Gas Company** (now called **Northwest Central Gas Company**, following the 1982 company take-over by **Northwest Pipeline Company**) offered the first significant outlet for this midcontinent gas. In 1927, a year after the big Hugoton discovery, the company laid 250 miles of 20-inch pipe to connect Wichita, Kansas. The following year, **Colorado Interstate Gas Company** opened up the field to consumers in Denver via a 450-mile transmission system. By

**Table 3-2.   A Chronology of Gas Pipeline Construction.**

| Date Completed | From | To | By | Total Miles | Pipe Diameter |
|---|---|---|---|---|---|
| | | BIRTH OF THE TRANSMISSION INDUSTRY | | | |
| 1872 | Titusville, PA | Titusville, PA | N/A | 5 | 2 |
| 1889 | Findlay, OH | Detroit, MI | N/A | 92 | N/A |
| 1891 | Greenstown, IN | Chicago, IL | N/A | 120 | 8 |
| 1909 | West Virginia | Cleveland, OH | N/A | 183 | 20 |
| 1909 | West Virginia | Cleveland, OH | N/A | 120 | 20 |
| 1909 | Northcentral Texas (Petrolia Field) | Dallas & Fort Worth, TX | Lone Star Gas Co. | 100 | 16 |
| 1911 | Northwest Louisiana | Little Rock, AR | Arkansas-Louisiana Gas Co. | 16 | 16–18 |
| 1913 | Southern California (Buena Vista Field) | Los Angeles, CA | N/A | 120 | 13 |
| | | THE LATE 1920s PIPELINE BOOM | | | |
| 1925 | Southern Texas | Houston, TX | Houston Pipe Line Co. | 220 | 12–18 |
| 1925 | Monroe Field (Northern Louisiana) | Beaumont, TX | Magnolia Petroleum Co. | 214 | 14–18 |
| 1925 | Monroe Field | Houston, TX | Dixie-Gulf Gas Co. | 217 | 22 |
| 1926 | Monroe Field | Baton Rouge, LA (90 mi. New Orleans, 1928) | Interstate Natural Gas Co. | 170 | 22 |
| 1927 | Panhandle Field (Northern Texas) | Wichita, KS | Cities Service Gas Co. (now Northwest Central Pipeline Corp.) | 250 | 20 |
| 1928 | Panhandle Field | Northcentral Texas (connection with pipes at Petrolia Field) | Lone Star Gas Co. | 200 | 18 |
| 1928 | Panhandle Field | Denver, CO | Colorado Interstate Gas Co. (then, Southwest Development Co.) | 350 | 20–22 |
| 1928 | Jennings Field (Southwest Texas) | Monterrey, Mexico | United Gas Co. & Mexican subsidiary | 141 | 18 |

| Year | Source | Destination | Company | Miles | Diameter (in.) |
|---|---|---|---|---|---|
| 1928 | Monroe Field | St. Louis | Mississippi River Fuel Corp. | 350 | 20–22 |
| 1929 | Permian Basin (Jal Field, NM) | El Paso, TX | El Paso Natural Gas Co. | 218 | 16 |
| 1929 | Monroe Field | Atlanta, GA | Southern Natural Gas Co. | 460 | 20–22 |
| 1929 | Monroe Field | Memphis | Texas Gas Transmission Co. (then, Memphis Natural Gas Co.) | 210 | 18 |
| 1929 | San Joaquin Valley (Kettleman Hills Field) | San Francisco | Pacific Gas & Electric Co. | 297 | 16–22 |
| 1929 | Southwest Wyoming | Salt Lake City, UT | Mountain Fuel Supply Co. | 290 | 14–18 |
| 1930 | Panhandle/Hugoton Field | Minneapolis via Omaha | Northern Natural Gas Co. | 1100 | 24–26 |
| 1931 | Panhandle/Hugoton Field | Chicago, IL | (now) Natural Gas Pipeline Co. of America | 980 | 24 |
| 1931 | Panhandle/Hugoton Field | Indianapolis, IN | Panhandle Eastern Pipe Line Co. | 900 | 20–24 |
| 1931 | Eastern Kentucky | Washington, D.C. | Columbia Gas & Electric Co. "Atlantic Seaboard Pipeline" | 467 | 20 |

THE GREAT DEPRESSION AND WORLD WAR II

| Year | Source | Destination | Company | Miles | Diameter (in.) |
|---|---|---|---|---|---|
| 1932 | El Paso, TX (from El Paso's pipe) | Southeast Arizona (and adjacent Mexico) | Western Gas Co. | 275 | N/A |
| 1933 | Central Michigan | Flint and Pontiac, MI | Michigan Natural Gas Co. | 157 | N/A |
| 1936 | Indianapolis (connecting with existing pipe) | Detroit, MI | Panhandle Eastern Pipe Line Co. | 300 | N/A |
| 1944 | Texas Gulf Coast | Cornwall, WV | Tennessee Gas Transmission Co. | 1265 | 24 |

THE POSTWAR PIPELINE BOOM

| Year | Source | Destination | Company | Miles | Diameter (in.) |
|---|---|---|---|---|---|
| 1945 | Monroe Field and Carthage Field, East TX | New York City | Texas Eastern Transmission Corp. (conversion of oil pipelines): "Big Inch" "Little Big Inch" | 1340 / 1475 | 24 / 20 |
| 1947 | Permian Basin | Southern California | El Paso Natural Gas Co. | 1200 | 26–30 |
| 1949 | Panhandle/Hugoton Field | Milwaukee, WI, and Detroit, MI | Michigan-Wisconsin Pipe Line Co. (subsidiary of American Natural Resources) | 1609 | 24 |

**Table 3-2** *(continued)*.   A Chronology of Gas Pipeline Construction.

| Date Completed | From | To | By | Total Miles | Pipe Diameter |
|---|---|---|---|---|---|
| 1950 | Texas-Louisiana Gulf Coast | New York City | Transcontinental Gas Pipe Line Co. | 1840 | 26-30 |
| 1950 | San Juan Basin, NM | San Francisco, CA | El Paso Gas Co. (NM to Topock, AZ) Pacific Gas & Electric Co. (Topock to San Francisco) | N/A | 24-34 |
| 1951 | Northwest Pennsylvania (connecting with Tennessee's existing pipe) | Northern Massachusetts | Tennessee Gas Transmission Co. | 520 | N/A |
| 1951 | Texas Gulf Coast | Chicago, IL | Natural Gas Pipeline Co. of America (then, Texas-Illinois Natural Gas Pipeline Co.) | 1300 | 26-30 |
| 1951 | Louisiana Gulf Coast | Central Illinois (connecting with Panhandle pipe) | Trunkline Gas Co. (now a subsidiary of Panhandle Eastern) | 1300 | 24-26 |
| 1953 | New Jersey (from Texas Eastern pipe) | Boston, MA (via Connecticut) | Algonquin Gas Transmission Co. (affiliate of Texas Eastern) | N/A | N/A |
| 1953 | Reversal of El Paso's line between Panhandle and western Texas to move Permian Basin gas north for the first time, connecting with Northern Natural | | | | |
| 1954 | Louisiana Gulf Coast | West Virginia | Gulf Interstate Company (now, Columbia Gulf Transmission Co.) | 1150 | 30 |
| 1954 | Louisiana Gulf Coast | Pipe terminus, Monroe Field | Southern Natural Gas Co. | 450 | 20-24 |
| 1956 | Louisiana Gulf Coast | Michigan | (now) Michigan-Wisconsin Pipe Line Co. | 1200 | 30 |
| 1956 | San Juan Basin | Pacific Northwest (Seattle, Portland, Spokane) | Pacific Northwest Pipe Line Co. (now, Northwest Pipe Line Co.) | 1487 800 | 22-26 smaller |

THE FOREIGN CONNECTION

| Date Completed | From | To | By | Total Miles | Pipe Diameter |
|---|---|---|---|---|---|
| 1957 | Alberta, Canada | Vancouver and U.S. border | Westcoast Transmission Co. (connecting with Pacific Pipe Line) | 650 | 30 |
| 1957 | Mexican border | Pipe connections in Eastern Texas with Midwestern Gas Transmission | Texas Eastern Transmission Corp. | 422 | 30 |

| Year | Origin | Destination | Company | Miles | Diameter |
|---|---|---|---|---|---|
| 1958 | Alberta, Canada | Montreal and U.S. connection with Midwestern Gas Transmission | TransCanada Pipelines | 2287 | various |
| 1959 | Southern Texas coast | Miami, FL | Houston Corp.     main:<br>Florida Gas Trans. Co.   supply:<br>distribution lateral: | 1517<br>414<br>682 | 24 |
| 1959 | Company pipeline terminus in Tennessee | Chicago, IL | Midwestern Gas Transmission Co.<br>(a subsidiary of Tennessee Gas Trans.) | 350 | 30 |
| 1960 | Manitoba/Minnesota border | Marshfield, WI | Midwestern Gas Transmission Co. | 504 | 24 |
| 1960 | San Juan Basin and Permian and Panhandle Fields | California border at Topock, AZ | Transwestern Pipeline Co. | 1300 | 24-30 |
| 1961 | Alberta ("Foothills" belt) | San Francisco, CA | Pacific Gas Transmission Co. | 1400 | 36 |
| 1967 | Manitoba/Minnesota border | Detroit, MI (and beyond into Ontario) | Great Lakes Gas Transmission Co.<br>(co-owned by Michigan-Wisconsin and TransCanada) | 989 | 36 |

**Map 3-2.  The Pipeline Boom of the Late 1920s.**

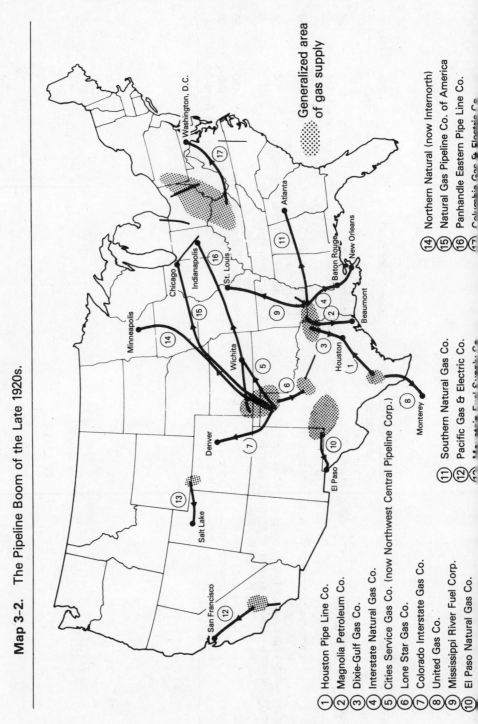

Generalized area of gas supply

1  Houston Pipe Line Co.
2  Magnolia Petroleum Co.
3  Dixie-Gulf Gas Co.
4  Interstate Natural Gas Co.
5  Cities Service Gas Co. (now Northwest Central Pipeline Corp.)
6  Lone Star Gas Co.
7  Colorado Interstate Gas Co.
8  United Gas Co.
9  Mississippi River Fuel Corp.
10  El Paso Natural Gas Co.
11  Southern Natural Gas Co.
12  Pacific Gas & Electric Co.
14  Northern Natural (now Internorth)
15  Natural Gas Pipeline Co. of America
16  Panhandle Eastern Pipe Line Co.
17  Columbia Gas & Electric Co.

**Table 3-3.**  The Major Gas-Producing Areas
in the United States.

| Area | Original Reserves[a] (trillion cubic feet) | Percentage of Total U.S. |
|---|---|---|
| Louisiana Gulf Coast | 169 | 23 |
| Texas Gulf Coast | 130 | 17 |
| Anadarko/Amarillo fields (Hugoton/Panhandle) | 117 | 16 |
| Permian Basin | 73 | 10 |
| East Texas (including Carthage field) | 50 | 7 |
| Appalachian region | 39 | 5 |
| California (all basins) | 33 | 4 |
| Prudhoe Bay field (Arctic Alaska) | 26 | 3 |
| San Juan Basin | 22 | 3 |
| Other U.S. | 92 | 12 |
| TOTAL U.S.[b] | 751 | 100 |

Source: Nehring, Richard (of the Rand Corporation, prepared for the U.S. Geological Survey), *The Discovery of Significant Oil and Gas Fields in the United States* (Washington, D.C.: U.S. Government Printing Office, January 1981).

[a]Estimate of the total amount of original proved recoverable identified through 1979.

[b]Of the 751 tcf of original reserves in place, 195 tcf have not yet been produced.

1931, three more pipelines—each stretching about a thousand miles—were in place, linking these vast reserves with consumers in the upper Midwest and the Great Lakes region. Also in 1931, Panhandle gas flowed into the eastern states for the first time, via a connection with Columbia's existing network.

A second locus of construction activity was the **Monroe** field of northern Louisiana, discovered in 1916. In 1928 and 1929, three distinct pipeline companies united this source of nonassociated gas (about 7 tcf) with consumers in St. Louis, Memphis, and Atlanta. Today, the Monroe field is nearly exhausted, with gas oozing out of old wells at about 20 psi.

Meanwhile, the **San Joaquin Valley** in central California was continuing to supply Los Angeles with gas from the Buena Vista field. Discovery of the larger Kettleman field (about 3 tcf of associated gas) in 1928 justified construction of a 300-mile pipeline to San Francisco.

The late 1920s pipeline boom spawned the first major gas transmission systems and companies, including many that are industry giants today: **Northern Natural Gas Company, Panhandle Eastern Pipe Line Company, Cities Service Gas Company** (now **Northwest Central Gas Company**), **Colorado Interstate Gas Company, Southern Natural Gas Company, Pacific Gas and Electric Company,** and **El Paso Natural Gas Company**. In addition, the trunk pipelines now owned by **Texas Gas Transmission Corporation** and **Natural Gas Pipeline Company of America** were laid during that era. Table 3–2 provides details on these and later pipeline systems.

## THE GREAT DEPRESSION AND WORLD WAR II

The Great Depression overtook even the burgeoning natural-gas industry. Construction of long-distance gas pipelines came to a standstill between 1932 and the United States' entry into World War II. The war stimulated energy consumption in East Coast industrial centers, but enemy submarines made it risky to move oil from western fields by tanker. Between December of 1941 and May of 1942, fifty-five domestic tankers were sunk off the U.S. coast. In response to this wartime fuel crisis, the federal government granted **Tennessee Gas Transmission Company** special privileges to procure labor and steel for a 1,275-mile gas pipeline. (See Map 3–3.) This system was the first to bring gas to the Appalachian region from the **Gulf Coast** of Texas and Louisiana.

The government itself built two National Defense oil pipelines, popularly called the Big Inch and the Little Big Inch pipelines. The two lines originated in the east Texas oil fields, bringing crude oil and refined products into the upper Midwest and the Appalachians. Immediately after the war, the Inch lines were converted for natural-gas transmission. Construction of the Tennessee system and conversion of the Inch lines marked a turning point in the history of long-distance gas transmission. Their stories are examined next.

### Construction of the Tennessee Gas Transmission System

**Tennessee Gas Transmission Company** (then, Tennessee Gas and Transportation Company) organized in the late 1930s to build a gas

**Map 3-3.** Pipeline Construction During the Great Depression.

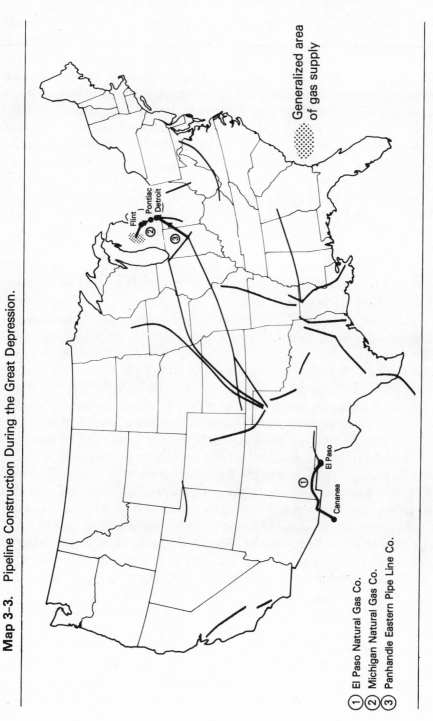

Generalized area of gas supply

Flint
Pontiac
Detroit

El Paso
Cananea

① El Paso Natural Gas Co.
② Michigan Natural Gas Co.
③ Panhandle Eastern Pipe Line Co.

pipeline between the Louisiana Gulf Coast and the Tennessee Valley. The company could not arrange financing, however, before steel rationing limited construction opportunities. Finally in 1943, the War Production Board (WPB) authorized the materials for construction of a single gas-transmission line to bring gas into the industrial East.

By this time, Tennessee had changed its strategy. Instead of targeting the Tennessee Valley, the company planned to sell gas to the regional distributor at Cornwall, West Virginia. Meanwhile, **Hope Natural Gas Company** announced its interest in building a line to the same consuming area but tapping Hugoton instead of South Louisiana reserves.

It was now up to the Federal Power Commission (FPC) to decide which of the two competing proposals would be certified to move ahead with construction. The WPB had voiced no preference, but it did make its approval of steel procurement valid only through September 1943. The board set this deadline to ensure that somebody ordered pipe while it was still possible to get the system functioning for the 1944–45 heating season.

The Federal Power Commission issued a conditional certificate to Tennessee Gas in July 1943. Final certification would follow when the company secured financing. By September, Tennessee Gas was still looking for construction capital, and the FPC announced that on September 21 it would hold hearings to consider Hope's application. On September 20, however, Tennessee's financing came through, and the commission granted final certification almost immediately.

Tennessee's achievement was a surprise. The company had more political than financial clout; it had not yet acquired any physical assets, but its presiding officer was the former son-in-law of President Roosevelt. Moreover, from time to time, Tennessee had suggested that this project might be an appropriate enterprise for federal financial aid.

Hope, however, looked like an obvious winner. As a subsidiary of the Standard Oil Company of New Jersey (now Exxon), its ability to raise capital was unquestioned. And while Tennessee was a newcomer to the transmission business, Hope Natural Gas was a well-established carrier in the Appalachian region. What is more, both companies intended to interconnect with Hope's existing system at Cornwall.

From the standpoint of reserves, too, Hope's proposal seemed superior. The Panhandle-Hugoton area was already connected to the Midwest by several large-diameter pipelines, whereas Gulf Coast gas had not yet been marketed outside of the immediate vicinity. Discoveries were so new that the existing pipelines between Houston and northern Louisiana still moved gas southward to the Texas coast from the older Monroe field.

Tennessee also had to overcome the vehement opposition of Texas and Louisiana officials. Recent industrial growth in these states depended on cheap energy in the form of both casinghead and non-associated gas for which no other markets existed. State officials feared that a gas-export market would thwart further industrial expansion. The governor of Louisiana, in fact, had made interstate acquisition of reserves so difficult that in late summer of 1943 Tennessee extended its proposed pipeline route beyond Louisiana into Texas, where political opposition, though strong, was less focused.

Despite all these points in Hope's favor, the Federal Power Commission selected Tennessee. The commission justified its choice on the grounds that Tennessee had satisfied all mandates of the conditional certificate and that the company had been trying to put together its project for a lot longer than Hope had shown an interest. The FPC also claimed that the deadlines set by the War Production Board simply did not allow full consideration of the competing application.

In its written Opinion on the matter, the commission ruled that it had no authority to determine end-uses of natural gas nor to arbitrate questions of interfuel competition. This ruling allowed it to issue the certificate without addressing either the arguments raised by officials in Texas and Louisiana or the generic objections that the coal industry routinely voiced in pipeline proceedings.

The Tennessee Gas pipeline was completed in 1944 at a cost of $4 million. (See Map 3–4.) Though Tennessee Gas and Transportation Company had emerged victorious, its sponsors had given up a lot in the process. Difficulties in putting together a financing package forced the owners to sell 90 percent of the common stock and all of the preferred stock to Chicago Company for $500,000. The Chicago Company (part of Sammuel Insull's financial empire and the predecessor of Peoples Gas Light and Coke Company) owned 2 tcf of gas reserves along the Texas Gulf Coast. Dedication of these reserves insured the new pipeline continuing supplies for twenty-six years at the intended throughput rate of 207 mmcf per day.

### Conversion of the "Inch" Oil Pipelines

During World War II, the U.S. government built two oil pipelines for the purpose of national defense. The **Big Inch** pipeline carried crude

**Map 3-4.** Pipeline Construction during World War II.

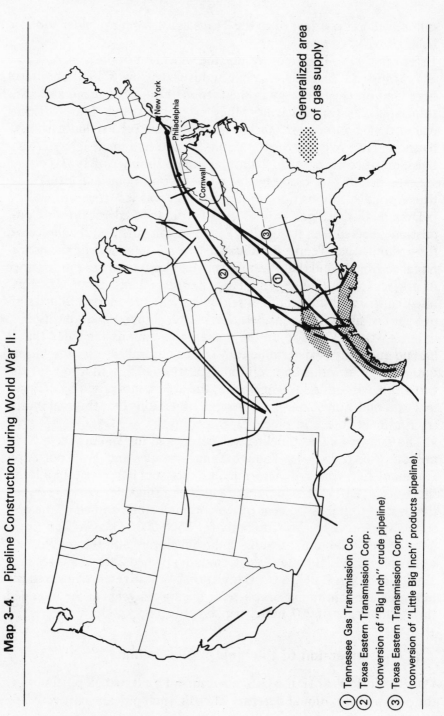

Generalized area of gas supply

New York
Philadelphia
Cornwall

① Tennessee Gas Transmission Co.
② Texas Eastern Transmission Corp.
    (conversion of "Big Inch" crude pipeline)
③ Texas Eastern Transmission Corp.
    (conversion of "Little Big Inch" products pipeline).

oil 1,340 miles from Longview, Texas, to Philadelphia. The **Little Big Inch** pipeline stretched 1,475 miles, bringing refined products from Beaumont, Texas, to New York City.

With the resumption of tanker activities in November 1945, the Inch lines fell idle. Congress created the War Assets Administration (WAA) to dispose of government properties accumulated during the war, including the Inch lines. Controversy raged over whether the WAA ought to consider the intended use of the pipelines when reviewing private bids.

Some influential people feared that an oil company would purchase the line and fill it with concrete, just to protect its own marketing system from competitors. Others argued that the government ought to ensure conversion of the Inch lines to natural gas because close to 14 percent (680 bcf) of the nation's annual gas production was still being flared. Not surprisingly, the coal industry suggested that the WAA ought to prohibit the new owners from moving natural gas through the Inch lines because it would endanger jobs in eastern coal fields. Those fears were well grounded. Coal gas was selling for about 70 to 90 cents per million btu (mmbtu), while industry analysts estimated that natural gas could be delivered through the Inch lines for only 24 to 26 cents per mmbtu.

Historical circumstances intervened to settle the controversy. A coal strike threatened eastern consumers who depended on manufactured gas. The federal government responded by activating the Inch lines for emergency deliveries of natural gas. Tennessee Gas Transmission was hired to make the necessary equipment conversions and received temporary approval to ship 150 mmcf daily into the energy-hungry East.

With gas transmission underway, the WAA offered the Inch lines for competitive sale in early 1947 and received six valid bids. Most of the interested companies proposed to keep at least one of the lines in the gas-transmission business. As expected, Tennessee Gas was one of the bidders, offering $11 million more than the appraised value of the combined systems. But a newly formed group of entrepreneurs was the winner. **Texas Eastern Transmission Corporation** tendered an offer of $143 million, which was just $3 million short of the original cost of construction. Texas Eastern maintained both pipelines as gas facilities until 1957, when the Little Inch was converted to carry refined oil products.

## THE POSTWAR PIPELINE BOOM

Rationing of consumer appliances and industrial steel ended along with the war, unleashing a frenzy of pipeline construction that lasted until the mid-1960s. The Panhandle and Hugoton fields, which had captured industry enthusiasm during the late 1920s pipeline boom, were still supporting new pipeline growth in the postwar era. Companies with established routes linking these midcontinent fields with the Midwest expanded their systems by installing more compressors and *looping* (double-piping) sections of the mainline. Today, **Northern Natural Gas Company, Panhandle Eastern Pipe Line Company**, and **Natural Gas Pipeline Company of America** all ship gas to the Midwest through systems that have been looped with parallel pipelines three or even four times.

After World War II, however, only one newcomer, **Michigan-Wisconsin Pipeline Company** (now a subsidiary of American Natural Resources Company), set up operations along the midcontinent-to-Midwest corridor. Attention turned, instead, to Texas. The **Carthage** gas field in eastern Texas (6 tcf of orignial reserves-in-place) had been discovered during the 1930s, but it was the rash of successful wildcats during the same decade along the **Texas Gulf Coast** (and later, along the Louisiana Gulf coast) that fueled the postwar boom.

The Gulf Coast region accounts for about 40 percent of all U.S. gas discovered to date. Between 1950 and 1956, five pipelines, each exceeding a thousand miles in length, were built from the Gulf Coast to points north and east. (See Map 3–5 and Table 3–2 for details.) The first company to complete its system was **Transcontinental Gas Pipe Line Corporation (Transco)**. An unsuccessful contender in the Inch Lines auction, Transco was determined to build its own pipeline along the same route from the Gulf Coast to New York City.

Meanwhile, producers in two western gas provinces found market outlets in California. Casinghead gas from the **Permian Basin** in west Texas and southeast New Mexico (now estimated to have contained 73 tcf initially) penetrated both northern and southern California via new and existing facilities owned by **El Paso Natural Gas Company**. Producers of nonassociated gas in the **San Juan Basin** of northwest New Mexico and adjacent Colorado (about 22 tcf) plugged into the San Francisco market via a joint construction effort by El Paso and one of California's two big **intrastate** gas carriers, **Pacific Gas and Electric Company**.

In addition to tapping these new reserves, pipeline construction in the early postwar era opened up new markets. By 1955, thousands of miles of pipeline connected small local gas fields with communities in Montana and Wyoming, extending as far east as Bismarck and Rapid

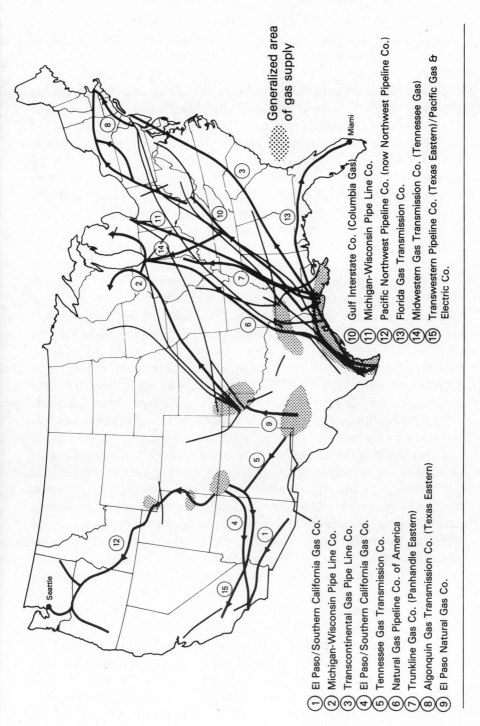

Generalized area of gas supply

Miami

Seattle

1. El Paso/Southern California Gas Co.
2. Michigan-Wisconsin Pipe Line Co.
3. Transcontinental Gas Pipe Line Co.
4. El Paso/Southern California Gas Co.
5. Tennessee Gas Transmission Co.
6. Natural Gas Pipeline Co. of America
7. Trunkline Gas Co. (Panhandle Eastern)
8. Algonquin Gas Transmission Co. (Texas Eastern)
9. El Paso Natural Gas Co.
10. Gulf Interstate Co. (Columbia Gas)
11. Michigan-Wisconsin Pipe Line Co.
12. Pacific Northwest Pipeline Co. (now Northwest Pipeline Co.)
13. Florida Gas Transmission Co.
14. Midwestern Gas Transmission Co. (Tennessee Gas)
15. Transwestern Pipeline Co. (Texas Eastern)/Pacific Gas & Electric Co.

City in the Dakotas. These small-diameter pipelines (8 to 16 inches) are not, however, depicted on Map 3–5, which shows only the 24- to 30-inch trunklines that characterized interstate construction during this same period.

One of the most promising frontiers in U.S. gas marketing during the early 1950s was New England. Subsidiaries of Tennessee Gas Transmission Company and Texas Eastern Corporation (**Algonquin Gas Transmission Company**) extended their transmission networks into this area in 1953, providing Gulf Coast producers with a huge new outlet. Federal certification of these two projects followed an unusually bitter contest between the applicants. Each wanted an exclusive franchise to serve the entire region, and old grudges stemming from Texas Eastern's victory over Tennessee Gas in the Inch Lines sale made it difficult for the two companies to strike an accord.

In considering the two applications, the Federal Power Commission had to cope with protests from eastern states already receiving natural gas. State leaders argued that it was unfair for transmission companies to open up wholly new markets when demand in existing markets still outstripped supplies, forcing distributors to restrain new hook-ups and to impose seasonal curtailments of deliveries to established customers. The FPC, however, was not moved by these pleadings. It authorized Texas Eastern to bring gas into Connecticut, Rhode Island, and eastern Massachusetts, while Tennessee got the rest of Massachusetts and upstate New York.

Throughout most of the postwar boom, the nation's three major producing areas each maintained distinct market outlets. Panhandle gas flowed into the Ohio Valley and Great Lakes region, gas from the Gulf Coast entered the Appalachians and the Eastern Seaboard, and Permian gas was routed west. In 1951, however, two systems were laid from the Gulf Coast directly north. **Trunkline Gas Company** (now a subsidiary of Panhandle Eastern Pipe Line Company) built a system that joined Panhandle Eastern's facilities in central Illinois, and another pipeline (which is now part of the **Natural Gas Pipeline Company of America**) carried Gulf Coast gas directly to Chicago.

In 1953, Permian gas began flowing north and east for the first time. Northern Natural pioneered this connection, reversing the flow of a pipeline that had formerly carried Panhandle gas southward into western Texas. Similarly, **Southern Natural Gas Company** was able to tap Gulf Coast resources in 1954 by building a relatively short spur to its existing facilities in the dwindling Monroe field of northern Louisiana.

The postwar boom was still in full swing during the late 1950s as interstate transmission companies augmented their systems by adding parallel lines, loops, and compressors to accommodate growth in existing markets. Only two new routes were established, however, and these penetrated the remaining virgin markets in the United States: the Pacific Northwest and Florida.

The Pacific Northwest drew gas from the San Juan Basin beginning in 1956. **Pacific Northwest Pipeline Company** staked out a challenging 1,500-mile path through basin-and-range topography. The pipeline also drew gas from the Rangely Field of Colorado and the Big Piney Field of Wyoming, and it stimulated gas exploration throughout the Rockies. Shortly after operations commenced, El Paso Company bought out practically all of the Pacific Northwest stock. The Justice Department's antitrust division opposed the transaction, contending, in part, that the take-over created a *monopsony* in which El Paso was the sole purchaser of gas from the San Juan Basin. The Department's arguments finally prevailed, and in 1974 El Paso began to divest itself of the Pacific Northwest holdings. The new enterprise, now called **Northwest Pipeline Company**, did not become wholly independent of El Paso until 1979.

This bold attempt by El Paso to increase its market power was preceded by a similar incident that put the FPC's certification policies to the test. In the late 1950s, a fledgling corporation, **Transwestern Pipeline Company** (now a wholly-owned subsidiary of Texas Eastern Gas Transmission Corp.), approached the FPC for approval of plans to build a pipeline between the San Juan Basin and California, adjacent to El Paso's pipeline. El Paso retaliated with a competing application, proposing to accommodate expected growth in California demand by simply looping its present system.

The elder company was able to put together more favorable supply arrangements than Transwestern, both from the standpoint of volume and price. But the Federal Power Commission was eager to break El Paso's monopoly in California. Transwestern was awarded the certificate, provided it renegotiated its supply contracts. The company was successful, and in 1960 gas began to flow through the new system. El Paso was, however, only a temporary loser. California demand-growth soon justified renewed pipeline expansion. Today, El Paso's northern system consists of three, and in places four, parallel lines between the San Juan Basin and the California border.

Gas came to Florida in 1959 when the Houston Corporation (now **Florida Gas Transmission Company**) built over 1,500 miles of mainline from the southern tip of Texas to Miami, Florida. Four hundred miles of spur pipelines tapped resources along the Gulf Coast, while almost 700 miles of laterals carried gas throughout Florida. Certification of the Florida pipeline was far from routine. The sponsors proposed to operate the system as a *contract carrier* in addition to the more usual status of *private carrier*. In those days, almost without exception, interstate pipelines carried only gas that they themselves had purchased in the field. Houston Corporation, however, intended that 60 percent of the pipeline's throughput would be handled on a contract basis for two big electric utilities in Florida. Those utilities had arranged purchase agreements with a number of producers along the Texas and Louisana coasts.

In late 1956, a bare majority of the Federal Power Commission issued Houston Corporation a certificate for pipeline construction, subject to several conditions. Among the most important was the stipulation that Houston Corporation tighten its transportation contracts with the two Florida utilities. The FPC wanted assurance that those buyers would not cancel purchases if delivered gas cost more than alternative fuels—most notably, *residual oil*, which the utilities were currently burning. Because of the semitropical climate, residential and commercial space-heating customers were far too limited and sporadic in their purchases to support a transmission system on their own. The FPC was, therefore, worried that the utility customers might walk away from their commitments, thereby shifting the entire burden of fixed transportation costs onto small-volume users. In the spring of 1957, Houston Corporation fulfilled the FPC requirements, and the commission granted final approval. That action, however, was challenged by the Florida Economic Council. It took another two years before the courts vanquished all questions about the legality of the certificate.

Until quite recently, a pipeline with a big industrial/utility load was considered rather fortunate. It did not have to cope with the huge swings in seasonal demand that confronted pipelines in heavily residential markets. In 1980, Florida Gas Transmission was still delivering over half of its gas on a contract basis to electric utilities, but the price increase occasioned by the 1978 Natural Gas Policy Act posed exceptional problems for the company. Industrial and electric utility customers are, after all, the most price-sensitive of all gas users because they will readily switch to alternative fuels if gas prices approach those of residual oil or coal. Florida Gas petitioned the Federal Energy Regulatory Commission (FERC) for the right to abandon its gas services and to convert the pipeline to oil transportation. In 1982 the request was granted, though contesting parties managed to stall implementation for a year.

## THE FOREIGN CONNECTION: 1957 to 1960

Since 1929, the United States had been exporting gas to industrial customers in Monterrey, Mexico through a U.S.-owned pipeline. Following World War II, the FPC authorized La Gloria Corporation to export additional gas from McAllen, Texas, to Monterrey via a Mexican-owned facility. Between 1957 and 1960, the United States for the first time began receiving volumes of gas (in excess of the trickle of gas traditionally supplied to border towns in Texas) from both of its continental neighbors.

Negotiations with **Petroleos Mexicanos S.A. (Pemex)** began in 1953 when several interstate pipeline companies approached the Mexican national petroleum company for sale of gas reserves discovered in 1945 just south of the Texas border. Two years later, **Texas Eastern Transmission Corporation** secured a contract to purchase up to 200 mmcf per day at an initial price of 14 cents per mcf (to be renegotiated every five years). But U.S. regulatory approval was delayed by protests from Texas producers. Those producers argued that Pemex's price was exorbitant considering that U.S. gas was currently flowing south to industrial customers in Monterrey for about a third as much.

Despite these protests, the Federal Power Commission authorized Texas Eastern to build 400 miles of 30-inch pipeline from the southernmost tip of Texas to the company's existing transmission network in the eastern part of the state. (See Map 3–6.) Deliveries of 150 mmcf per day began in 1958 and continued under the 1955 contract for a total of twenty years. Because the delivered volume was at the seller's discretion (subject to Mexico's internal needs), flow rates varied enormously. By 1964, the pipeline carried only a trickle of Mexican gas. (See Table 3–4.) The year 1957 also marked the onset of Canadian gas deliveries. British Columbia's **Westcoast Transmission Company** completed 650 miles of 30-inch trunkline, connecting population centers in the southwestern corner of the province with the Peace River Gas Field of

**Map 3-6.** The Foreign Connection: 1957 to 1960.

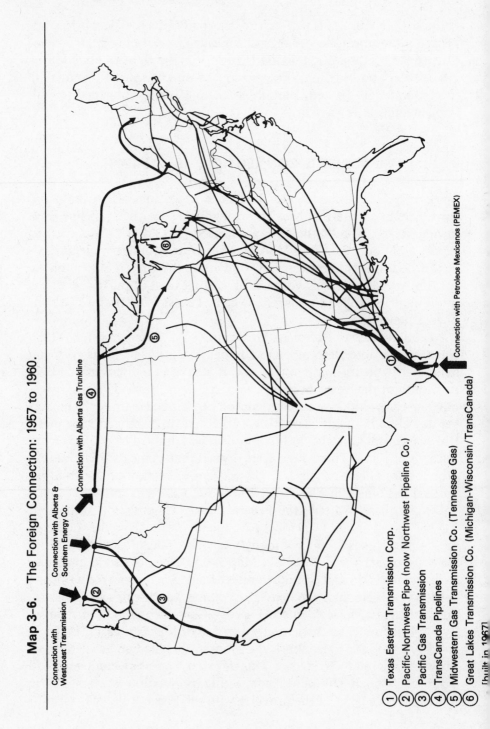

Connection with
Westcoast Transmission

Connection with Alberta &
Southern Energy Co.

Connection with Alberta Gas Trunkline

Connection with Petroleos Mexicanos (PEMEX)

① Texas Eastern Transmission Corp.
② Pacific-Northwest Pipe (now Northwest Pipeline Co.)
③ Pacific Gas Transmission
④ TransCanada Pipelines
⑤ Midwestern Gas Transmission Co. (Tennessee Gas)
⑥ Great Lakes Transmission Co. (Michigan-Wisconsin/TransCanada)
[built in 1967]

**Table 3-4.**   History of Gas Trade between the
United States and Mexico.

| Year | Exports to Mexico (billions of cubic feet annually) | Imports from Mexico | Net |
|------|-----------------------------------------------------|---------------------|-----|
| 1955 | 19.9 | negligible | 19.9 EX |
| 1960 | 10.5 | 47.0 | 36.5 IM |
| 1965 | 9.5 | 52.0 | 42.5 IM |
| 1970 | 14.7 | 41.3 | 26.6 IM |
| 1971 | 15.8 | 20.7 | 4.9 IM |
| 1972 | 14.6 | 8.1 | 6.5 EX |
| 1973 | 14.0 | 1.6 | 12.4 EX |
| 1974 | 13.3 | 0.2 | 13.1 EX |
| 1975 | 9.5 | 0.0 | 9.5 EX |
| 1976 | 7.4 | 0.0 | 7.4 EX |
| 1977 | 3.9 | 2.4 | 1.5 EX |
| 1978 | 4.0 | 0.0 | 4.0 EX |
| 1979 | 4.3 | 0.0 | 4.3 EX |
| 1980 | 3.9 | 102.4 | 98.5 IM |
| 1981 | 3.4 | 105.0 | 101.6 IM |

Source: American Gas Association, *Gas Facts, 1981* (Arlington, Va.: American Gas Association, 1982).

northern British Columbia and Alberta. Westcoast's system reached the international border at Sumas, Washington, where provincial exports were picked up by Pacific Northwest Pipeline Company and carried south. Canadian imports thus became an important source of energy for northwestern states only a year after the area was first served by gas from the San Juan Basin.

California received its first delivery of Canadian gas in 1961, by way of a 1,400 mile pipeline system. The U.S. portion was built by **Pacific Gas Transmission Company**, an interstate subsidiary of northern California's intrastate carrier, Pacific Gas and Electric Company (PG&E). PG&E also participated in building the British Columbia portion of the Canadian system (Alberta and Southern Gas Co., Ltd.), which originated in the petroleum-rich regions of western Alberta.

At about the same time, the United States began receiving Canadian gas via a third outlet—this time, connecting with eastern markets. The **TransCanada Pipeline** system, linking Alberta gas fields with Canadian and U.S. consumers as far east as Montreal, was probably the most ambitious North American pipeline construction ever. Almost 2,300 miles of mainline pipe spanned southern Canada. In addition, a 350-mile spur owned by an affiliate of Tennessee Gas Transmission Company (**Midwestern Gas Transmission Company**) was an integral part of the system, opening up gas markets in northern Minnesota and Wisconsin.

The choice of route for the TransCanada project was highly political. Company officials and prospective lenders preferred a southern approach that dipped into Michigan on the way to Montreal, thus avoiding the muskeg swamps and bedrock barrens north of Lake Superior. A decade later, in fact, TransCanada and Michigan-Wisconsin Pipeline Company jointly completed the **Great Lakes Gas Transmission** system, utilizing this more practical southern corridor. But in the late 1950s, political realities necessitated an all-Canadian line.

The project was so massive and the terrain so difficult that the sponsors had to take extraordinary measures to eliminate many of the usual risks associated with new construction. For example, because energy users in Montreal would have to be convinced to switch to gas, the TransCanada team devised a marketing approach whereby U.S. gas would temporarily flow into eastern Canada via a link with the eastern transmission network. U.S. gas would then become the basis for market development, but imports would terminate as soon as the all-Canadian system was in place. This scheme, however, required licenses from the U.S. Federal Power Commision, as well as from Canadian authorities. The FPC was exceedingly slow in reaching a decision to allow gas exports, despite intense diplomatic efforts by the Canadian government.

U.S. approval of Midwestern's spur into Minnesota and Wisconsin was another key component of project success. Juxtaposing construction of this U.S. spur with construction of the western portion of the TransCanada pipeline would allow the system to begin gas deliveries in advance of overall project completion. The revenues generated by the western portion of the project could then be used to strengthen the financial basis for construction of the eastern, all-Canadian line to Montreal.

But even these two ingenious strategies were not enough to induce banks and insurance companies to buy TransCanada's bonds. After six years, the sponsors (actually, several successive groups of sponsors) and the Canadian government finally decided that government support was unavoidable. The government agreed to provide financial backing for the riskiest pipeline segment, which traversed western Ontario. Federal financing, however, required Parliamentary approval, leading to one of the bitterest conflicts in Canadian political history. In the end, TransCanada won; but the majority party that had put the deal together was so damaged by the struggle that it lost the next election.

William Kilbourne, in his book *Pipeline* (1970), chronicles Trans-Canada's turbulent history. Readers in the 1980s will find striking parallels between TransCanada's genesis and the difficulties confronting U.S. and Canadian sponsors of the proposed Alaska Highway gas pipeline.

## SYSTEM MATURATION DURING THE 1960s

By the mid-1960s, the postwar pipeline boom was over. The transcontinental construction ventures of the preceding twenty years were succeeded by a variety of less dramatic projects that brought the U.S. gas transmission network to maturity. Spurs were built from the major trunklines to service new areas; trunklines were looped and compressor stations added. By 1966, natural gas was available to consumers in all of the forty-eight conterminous states. Today, the nationwide network is so interconnected that if direction of flow and institutional barriers were of no concern, practically anybody who sells gas could send it to anybody who buys gas, regardless of location.

Pipeline investment opportunities were noticeably on the decline by 1960. Transmission companies funneled their resources into distribution subsidiaries and, to a lesser extent, into production ventures. Diversification into new lines of business was also popular.

System maturity was not, however, the only reason for the downturn. The incoming Kennedy administration effected huge changes in federal regulatory policies. A more consumer-oriented Federal Power Commission monitored profits and looked for ways to trim the purported fat off pipeline tariffs. These actions prompted Mutual Fund companies to dump interstate pipeline stocks from their portfolios between 1960 and 1962.

In addition, the commission actively sought congressional amendments to strengthen its oversight of utility securities, its control of gas imports and exports, and its influence over pipeline interconnections. The transmission industry did not take kindly to any of these proposals; but most disconcerting was the (unsuccessful) attempt to extend federal jurisdiction into the realm of *direct sales* to industrial and electric-utility customers.

Behavior of several interstate pipelines had almost forced the commission to act. Pipeline companies applied for certification of major construction projects keyed to large-volume, *nonjurisdictional* users who would purchase gas directly from the pipeline rather than through a local distributor. By selling most or all of the capacity directly to large electric utilities and industries, the pipeline circumvented federal rate regulation, which applied only to "sales for resale." If the pipeline went one step further and operated strictly as a contract carrier, producers could enter into sales directly with end-users and thereby hope to escape wellhead price regulation altogether.

The commissioners worried that construction of new pipelines under such contractual arrangements ultimately would reduce the amount of gas available to residential customers, who, necessarily, depend on local distributors. The new Florida gas pipeline, approved by an Eisenhower-appointed commission, provided a real-world example of just such a system. As a result, the commission delayed approving, or outright disapproved, most of the new interstate pipeline proposals submitted in the early 1960s.

Among these was the **Oklahoma-Illinois Gas Pipeline**, sponsored by producers in the Anadarko Basin (Hugoton Field) who had contracted with industrial users in St. Louis. A joint proposal by El Paso and Colorado Interstate to connect Wyoming reserves with southern California was another major project that never broke ground. Its defeat turned, in part, on a lack of sufficient reserves, coupled with conservative FPC projections of California demand that argued for restraint in capacity additions. In the early 1980s, this plan was revitalized as significant volumes of *deep* gas were discovered in the Overthrust Belt of the Rocky Mountain states.

The biggest failure of the era, however, was the proposal by Tennessee Gas Transmission Company to service two electric utilities in Los Angeles with gas from the Texas Gulf Coast and Mexico. Tennessee pursued this project for most of the decade, during which time it shifted the route, adopted a new name (**Gulf Pacific Pipeline**), and managed to fend off several competing proposals. The company's tenacity in promoting this particular venture grew out of an aspiration to enter the western pipeline business, the growing incentive for producers to escape wellhead regulation, California's battle against air pollution, and the inability of electric utilities in Los Angeles to secure *firm* (noninterruptible) gas from their existing suppliers.

Overall then, the transmission industry found itself in a relatively stagnant period during the 1960s. Moreover, the onset of wellhead regulation and the increasing advantages of direct industrial sales spurred an interest in construction projects that failed to meet the standards of "public convenience and necessity" as perceived by a Federal Power Commission now greatly concerned with the interest of residential gas consumers.

More insidious, however, was the toll that wellhead price regulation was taking on the commitment of new gas supplies to the interstate pipeline network. Not only were exploration incentives muted by federal price controls, but companies that did discover gas (usually in their quest for oil), found *intrastate* purchasers decidedly more attractive. Eventually, of course, the depletion in reserves precipitated a *deliverability* crunch. During the 1970s the nation's major transmission companies and distributors in nonproducing states had to refuse new customers and to curtail deliveries seasonally to existing customers. Chapter 5 takes a detailed look at the regulatory roots of the gas shortages. But first, we will explore how the gas transmission industry coped with this most difficult decade.

## REFERENCES

Kilbourne, William. *Pipeline*. Toronto: Clarke, Irwin, and Company, Ltd., 1970.

Nehring, Richard. *The Discovery of Significant Oil and Gas Fields in the United States*. Washington, D.C.: U.S. Government Printing Office, 1981.

# 4 GAS SHORTAGES OF THE 1970s
## The Era of "Supplemental" Gas Boondoggles

## THE SPECTER OF A VANISHING RATE BASE

When government fixes the price of a commodity below market value, it artificially stimulates consumption, discourages new supplies, and thereby makes a shortage inevitable. In 1954 the Supreme Court ordered the Federal Power Commission to begin regulating wellhead prices of gas sold into interstate commerce. A surplus of gas reserves attributable to oil exploration and development, however, delayed the appearance of a shortage until the early 1970s. The ensuing gas supply difficulties turned into full-scale crises during exceptionally cold winters in the East and Midwest, and customers who were forced to convert to other fuels faced financial hardships brought on by the oil-price shocks and tight air-quality standards.

By mid-decade, the shortage (and particularly its uneven incidence across the nation) had become politically intolerable. In the East and the Great Lakes states, schools and factories closed down for weeks at a time, and even homeowners began to worry about freezing. To the people who were affected, and to their elected representatives, it seemed unjust that large industrial plants and electric utilities in Texas and Louisiana were still gulping down gas without restraint. An obvious culprit was the unequal regulatory treatment of inter- and intrastate gas sales. Chapter 5 surveys the governmental response; here we will explore how the shortages affected gas industry activities and investment decisions.

Once the market value of gas clearly exceeded the federal price ceilings, interstate pipelines were no longer able to acquire new domestic reserves from onshore wells. Instead, transmission companies and distributors in the producing states (like Texas and Louisiana) were able to outbid their interstate counterparts and still offer customers a source of energy at an attractive price. With the end of the postwar construction boom, and with virtually no hope of maintaining (much less increasing) their gas supplies, the interstate gas-transmission industry faced a bleak future.

The prospect of declining business is a profound challenge for any enterprise or industry, but the pipeline companies confronted even more formidable hardships than unregulated firms would have encountered under similar circumstances. A regulated utility may bill its customers for the full cost of serving them, including a reasonable profit; but most regulatory bodies (including the Federal Power Commission) allow a profit to be earned only on the *depreciated original cost* of a company's investment in facilities. Absent new construction, therefore, a pipeline company's *rate base* (and thus its profits) will diminish through time, even if it suffers no decline in gas sales.

The financial outlook of a regulated firm is not inherently threatened by this *vanishing rate base*. Capital recovered steadily over a twelve- to twenty-five-year period in accordance with approved *depreciation schedules* can be reinvested in profitable new ventures— regulated or unregulated. The situation immediately after World War II, for example, was ideal for interstate gas transmission companies because there was no shortage of prospects for new pipeline construction. During the 1970s, however, a now-mature pipeline network and faltering gas supplies combined to limit investment opportunities within the transmission industry itself. In order to perpetuate themselves as profitmaking enterprises, pipeline companies had to expand into unrelated lines of business.

## Diversification Strategies and Investment in Gas Storage

Diversification became a key investment strategy for interstate transmission firms during the 1970s, although the Public Utility Holding

Company Act of 1935 (see Chapter 8) constrained their ability to become multi-industry conglomerates. Many companies established or acquired subsidiaries in related but unregulated enterprises, like coal mining or petrochemicals. Some moved into totally new areas like land development, packaging, and hardrock mining. Nevertheless, the gas shortage itself opened up possibilities for investment in regulated ventures that shared important features with cross-country pipelining.

For example, some transmission companies (as well as gas distributors who were likewise faced with vanishing rate bases) invested in underground gas storage, utilizing spent gas fields or salt domes. The costs of storage proved to be relatively high (often doubling the per-unit cost of gas). The resulting improvement in winter peak-shaving capacity, however, allowed distributors to hook up additional residential and commercial customers—whose demand was seasonal but who were willing to pay higher prices for gas than were industrial concerns. Wellhead prices were still so far below the value of gas to these *high-priority customers* that the benefits of *load upgrading* seemed to justify its added expense.

### Construction of Offshore Pipelines

Gas-gathering systems on the federal Outer Continental Shelf (OCS) offered another important avenue for company growth. Since OCS gas is legally outside the boundaries of the adjacent state, there is no way to move it ashore without subjecting it to regulation as an interstate commodity. OCS gas, therefore, accounted for almost all of the added supplies that interstate companies were able to harness in the 1970s. Moreover, interstate transmission companies built and owned the big offshore gathering systems because producer-financing of such facilities offered two equally unattractive possibilities.

If a producer invested in an expensive gathering system, there was no assurance that the FPC would allow those production-related costs to be added onto the wellhead price ceiling. Alternatively, if the lines were treated as transmission plant, the owners would be stepping into the regulated utility sphere. Perhaps no group of companies is more reluctant to submit to regulatory oversight than petroleum producers, who prefer to take big financial risks in pursuit of big payoffs.

But there was no need for producers to select either of these disagreeable options for getting their gas to market. Interstate transmission companies jumped at the opportunity to bolster their gas supplies and at the same time expand their rate base by building offshore gathering systems. For example, five of the biggest pipeline companies joined ranks to build the High Island Offshore System. Also tapping Gulf Coast reserves was the Sea Robin system, finally put into operation in 1980. Sea Robin bolstered pipeline rate base by $228 million, and the two partners each had rights to half of the gas. Offshore California provided additional construction opportunities, most notably, the $100 million Pacific Offshore Pipeline.

## The Lure of Supplemental Gas

Even more than large-volume underground storage projects and offshore pipelines, the enterprises that captivated transmission industry attention during the 1970s were *supplemental gas* ventures. Supplementals fell into three categories: liquefied natural gas tankers and terminals, high-btu coal gasification, and pipelines from Alaska and the Canadian Arctic. The unifying feature of all these ventures was an extremely high investment requirement, seemingly justified by the nation's critical need to supplement its energy supplies.

High-cost gas was not unknown before the 1970s. For some time, transmission and distribution companies had funneled capital into *peak-shaving* facilities to accommodate seasonal peaks in demand. Investments included underground storage, *liquefied natural gas (LNG)* storage, and *substitute natural gas (SNG)* plants. These latter facilities converted light hydrocarbon liquids (principally naphtha) into vapors that could augment natural supplies of methane. By the mid-1970s, however, federal policymakers began to actively discourage SNG investments because plant owners competed with gasoline refiners for limited supplies of naphtha feedstocks. In 1981, underground storage capacity in the United States totalled 7.5 tcf; LNG storage could accommodate another 62 bcf, and sixteen SNG plants were operational.

What distinguished supplemental gas projects from other high-cost utility ventures was their *base-load* orientation. Sponsoring companies planned to generate supplemental gas year-round. Supplemental ventures were also noted for their extraordinary investment requirements, often in the billion-dollar range. The financial scale certainly promised to vanquish any residual fear a company might have with respect to vanishing rate base, but the need to attract megadollars, especially in light of a host of unusual risks posed by the projects, turned most supplemental gas ventures into pipe dreams.

## LIQUEFIED NATURAL GAS

The supplemental projects that made the greatest progress during the 1970s targeted liquefied natural gas. Although base-load LNG was unknown in the United States before mid-decade, small-scale LNG storage was an established practice.

In 1941, East Ohio Gas Company built the first commercial LNG plant in the United States. The Cleveland facility was designed to store in liquid form excess supplies of gas delivered in the summer for peak-shaving use during the winter. LNG storage was practical because methane chilled below its boiling point ($-258.7$ degrees Fahrenheit, or $-161.5$ degrees Centigrade) occupies only $1/625$ of the space that it requires at ambient temperatures and pressures.

In 1944, the Cleveland facility exploded, killing 130 people and destroying a big chunk of urban real estate. That disaster brought a halt to the domestic LNG storage business for almost twenty years. Safety improved, however, with the progress of technology, and by 1982 over fifty LNG peak-shaving plants were operating in the United States.

The mid-1960s witnessed not only a resumption of U.S. LNG storage but the beginnings of international LNG transportation. Oil-producing regions in the Middle East, Africa, Latin America, and Indonesia were flaring huge volumes of associated gas, while reserves of nonassociated gas in these same areas were simply *shut-in* for lack of a market. (See Table 4–1.) Technologies responded, and by the end of the decade, superinsulated *cryogenic* tankers carried Algerian gas to England and France and gas from southern Alaska to Japan.

During the 1970s, Libya, Brunei, Abu Dhabi, and Indonesia joined the ranks of LNG exporters, although Algeria retained its preeminence in the market. Spain, Italy, and the United States joined England, France, and Japan as LNG importers. Japan was (and is likely to remain throughout the 1980s) the predominant LNG purchaser. In 1981, it received half of the 1.6 tcf of LNG traded worldwide.

Construction in 1969 of facilities in southern Alaska marked the first base-load LNG plant in the United States, but this project differed

**Table 4-1.**    Flaring of Natural Gas in OPEC Nations, 1979.

| Nation | Billion Cubic Feet | Percentage Flared[a] |
|--------|--------------------|----------------------|
| Abu Dhabi | 293 | 60.9 |
| Algeria | 353 | 23.0 |
| Ecuador | 14 | 91.7 |
| Gabon | 64 | 92.2 |
| Indonesia | 222 | 22.5 |
| Iran | 558 | 39.8 |
| Iraq | 431 | 84.5 |
| Kuwait | 127 | 27.4 |
| Libya | 166 | 20.0 |
| Nigeria | 1,014 | 95.4 |
| Qatar | 81 | 34.7 |
| Saudi Arabia | 1,342 | 75.1 |
| Venezuela | 81 | 6.3 |
| Total | 4,746 | 44.4 |

Source: Organization of the Petroleum Exporting Countries, *Annual Statistical Bulletin 1979.*
[a]Percentage of gross production flared in 1979.

starkly from the domestic projects that followed. The Phillips-Marathon plant exported rather than imported natural gas.

Union Oil Company, Phillips Petroleum Company, and Marathon Oil Company owned large reserves of gas in Alaska's Cook Inlet. Local markets, however, were insignificant compared to the available resource. Union decided to market its share by processing gas into ammonia and urea which, as liquids, were easily moved via tanker. Meanwhile, Phillips and Marathon reached an agreement with two Japanese gas utilities and built a liquefaction plant and two tankers for shipment of LNG. Trade commenced in 1969 under a fifteen-year agreement. In late 1982, the U.S. government granted a five-year renewal of the Phillips-Marathon export license, at established plant-intake levels of close to 70 bcf annually (yielding 50 to 55 bcf of exports).

The nation's first regasification terminal designed to handle strictly imported LNG began receiving shipments from Algeria in 1971. It was not, however, a supplemental gas project. Distrigas Corporation had built its Boston terminal and storage facilities for peak-shaving purposes, rather than to achieve a supplemental year-round supply. Moreover, the construction investment (only $10 million) was of a modest scale because the most expensive piece of the operation, the tanker, was owned by a foreign firm. In 1981, Distrigas LNG accounted for less than 40 bcf, but it represented the only oceanic gas imports received anywhere in the United States.

### The Beginnings of Base-load LNG Imports

The first capital-intensive, base-load LNG venture that was of the supplemental breed was the Algeria I project sponsored by El Paso Natural Gas Company. Though a supply contract averaging 1 bcf per day (365 bcf per year) was signed with Algeria in 1969, regulatory, financing, and construction obstacles delayed project start-up for nine years, and a diplomatic stalemate sent it into early retirement two years later.

Interestingly, El Paso stood to gain not a single btu from this project. The only tangible benefit was corporate expansion into a new line of business that, while technically beyond the reach of U.S. rate regulation, seemed to offer much of the business security enjoyed by utility ventures. El Paso planned to build nine LNG tankers at a combined cost of about $1 billion.

The gas loaded at the Algerian liquefaction plant (owned by Sonatrach, the state oil company), would be regasified in Chesapeake Bay and Savannah, Georgia—nowhere near El Paso's transmission network in the southwestern states. Instead, Columbia Gas Transmission Corporation and Consolidated Natural Gas Company committed to purchase the LNG at the Maryland terminal, and Southern Natural Gas Company signed up for the Georgia deliveries. This arrangement offered the three purchasing companies a chance not only to acquire gas but to boost their own rate bases by hundreds of millions of dollars invested in the regasification plants.

Putting together this kind of money, however, was not easy. Gas-transmission companies, as regulated utilities, were never floating in spare cash. Consequently, they needed to *leverage* their equity contributions with a big infusion of debt capital. In contrast to overall *capital structures*, which typically held company debt to a 50 or 60 percent level, sponsors of supplemental gas projects planned debt financing to cover 75 percent and sometimes even 90 percent of total capital needs.

Such leveraging is difficult to achieve even when a project presents a low level of risk, but base-load LNG is far from risk-free. If an overseas liquefaction plant shuts off supplies, for example, or if the exporting nation demands unrealistic changes in the sales arrangements, LNG-tanker owners have few other options for replenishing cargo. Likewise, the highly pressurized gas can only be off-loaded at ports equipped with specialized gear. Small physical problems, too, can escalate into major disasters at LNG terminals, disrupting transactions for months or even years.

Fear of LNG port disasters generated a good deal of public inquiry and debate about facility siting, specifically, the degree to which local or state concerns should be allowed to influence or even prevent the location of LNG receiving terminals deemed to be in the national interest. California, in particular, was swept into the debate. Its two biggest gas distributors (Pacific Gas & Electric Company and Pacific Lighting Corporation) had to fight a long and costly battle to secure the necessary siting permits for regasifying LNG from Alaska and Indonesia. It was not until 1982, when the Federal Energy Regulatory Commission (FERC) ruled for a second time (on a court remand) that earthquake hazards were insufficient to warrant project disapproval, that the Western LNG sponsors (sometimes called Pac-Alaska and Pac-Indonesia) cleared all the federal siting obstacles.

By the time the state was scheduled to respond, the financing and marketing outlook had deteriorated to the point that the project sponsors opted for delay. Both applied to the California Public Utilities Commission for formal classification of the project as a "plant held for future use." That designation would permit the companies to include about $400 million of promotional and design expenses in their utility rate bases, thereby ensuring capital recovery and a return on investment—whether or not the facilities are ever built.

## Nonrecourse Project Financing

Although $400 million is a lot of money, it is far less than what El Paso, Columbia, Consolidated, and Southern invested in actual plant construction of the Algeria I project. The multibillion-dollar scale of capital requirements made it essential for these companies to secure *project financing* on a *nonrecourse* basis. In addition to *balance-sheet financing* (wherein debt is secured first and fully by the credit-strength of the company itself), many industries utilize *conventional* forms of project financing. Here, revenues from a particular project are pledged for repayment of debt principal and interest, backed secondarily by the general assets of the sponsors themselves.

Regarding the big LNG ventures, however, it would have been imprudent for any of the pipeline-company sponsors to put their own treasuries at risk if something went wrong and the project's revenues were not sufficient to meet scheduled debt payments. Moreover, many of their existing bondholders held *covenants* that put an absolute cap on the amount of additional debt the company could take on. But if the sponsors were unwilling to secure the debt for construction of the tankers and the regasification facilities, who would?

The Algeria I project used a combination of approaches. For the six tankers built in the United States, El Paso raised 75 percent of the financing in the form of U.S. bonds, guaranteed under Title 11 of the Merchant Marine Act of 1936. Columbia, Consolidated, and Southern formed separate project entities for their regasification facilities and committed to pay these new ventures a *minimum bill tariff* under "all events," including the possibility that gas deliveries might temporarily or permanently be curtailed for reasons beyond the sponsors' control. This latter doctrine of *force majeure* (French for "superior force") is a standard provision in most forms of business contracts. It evolved from a common-law tradition whereby contractual obligations were refutable if a party was unable to perform because of war, weather, or "an act of God."

The minimum-bill clause protected all debt service obligations, ensuring scheduled payments of both principal and interest. The sponsor's equity investments were, however, only partially protected. Minimum billing guaranteed return *of* equity but allowed no profit or return *on* equity (which is the owner's equivalent of interest charges).

The transmission companies were willing to bind themselves into minimum bill arrangements with their newly created project entities because they, in turn, could pass that obligation downstream to industrial and distributor customers. Initially, federal regulators chose to ensure that downstream customers had a choice in LNG purchases. In 1972, the commission awarded a certificate for facility construction that required *incremental pricing* of the regasified LNG. El Paso protested that the tariff mandate would preclude debt financing because downstream marketing would not be fully ensured. Pursuant to a court remand, the commission in 1977 acquiesced, allowing the pipeline buyers to use the same *rolled-in pricing* structure that is standard business practice throughout the industry.

Flow-through of LNG costs via rolled-in pricing was especially favorable to the project sponsors because most interstate transmission companies added *purchased gas adjustment (PGA)* clauses to their tariffs during the 1970s. PGA clauses allow a company to respond to inflating gas costs by boosting rates to downstream customers automatically (usually on a semiannual basis), without first petitioning for regulatory approval. PGAs, authorized by the FPC in 1972, were vital during the recurrent supply crises, especially in conjunction with other emergency programs that made it possible for interstate transmission companies to secure additional, but costly, gas supplies. Absent PGAs, the sheer volume of these transactions would have swamped the regulatory process. From the standpoint of supplemental gas projects, the existence of PGAs for ensuring downstream transfer of costs was a fortunate coincidence.

During the early 1980s, however, PGAs came under increasing attack. Consumers, distributors, and state regulatory commissions criticized the way that PGAs enabled interstate transmission companies to pass through what they considered to be imprudent purchases of *exempt* categories of gas under the Natural Gas Policy Act of 1978. Because price adjustments were automatic, downstream buyers had no opportunity through the hearings process to protest high-priced acquisitions.

Michigan voters approved an initiative and a referendum on the 1982 election ballot that spelled the death of PGAs in that state. Although the attack on PGAs at the distributor level is spreading, success ultimately depends upon similar actions by federal authority. In early 1983, several bills (including a bill sponsored by President Reagan) were introduced in Congress that would have seriously limited future use of PGAs.

The PGA war is reminiscent of the campaign launched in the late 1950s by distributors seeking veto power over acquisitions made by their pipeline suppliers. Those distributors garnered a temporary victory, but the impracticalities of obtaining the concurrence of all end-users prior to purchase commitments moved the Supreme Court to order a return to the status quo (*United Gas Pipe Line Co. v. Memphis Light, Gas and Water Division*, 358 U.S. 103 (1959).

## A Legacy of Failure

The minimum-bill tariffs and PGA clauses associated with the Algeria I project were put to the test in 1980. Shortly after deliveries commenced in 1978, Sonatrach gave notice that it had second thoughts about the fairness of the pricing provisions in its supply contract. In May 1979, the parties tentatively reached a new agreement, subject only to final approval by the governing bodies in both the supplying and receiving countries.

U.S. approval of the new $1.95-per-mmbtu price (*f.o.b.* Algeria) was granted in December 1979, but Algeria announced shortly thereafter that it believed a further increase was warranted. Algeria was seeking a gas price that would achieve parity with its crude-oil exports, irrespective of the fact that tankering, regasification, and other downstream charges would make it impossible for the LNG to compete with refined oil at the *burner tip*. The gas might still be marketable, however, because on both a physical and an accounting basis it could be blended with cheap gas that made up the bulk of supplies sold by U.S. gas companies.

The U.S. government reacted forcefully to Algeria's ever-increasing demands, fearing in part the signal that acceptance of such terms would send to Canada and Mexico, who kept edging toward higher and higher prices for their own gas exports. The Department of Energy prohibited El Paso from concurring in Algeria's price amendments, so in April 1980, Algeria cut off its deliveries.

The losses were stupendous. The three U.S. tankers in operation and the three still in the shipyards forced the U.S. government to make good on its loan guarantee of over $400 million. El Paso's remaining obligations on the U.S. ships and on its three French vessels, plus its own equity contributions, generated a gross loss of $547 million, of which a net loss of $365 million showed up in El Paso's accounts in 1980. To add insult to injury, the shipbuilders operating under a fixed-price contract sued El Paso in 1980 for $92 million in cost overruns. El Paso countersued for $114 million in damages to cover alleged faulty workmanship and the hardships incurred by late delivery of the tankers.

Money was not the only thing that El Paso lost in its LNG fiasco. Corporate independence was another. The LNG write-off (followed by a declining market in California and a nationwide gas glut) was a key factor prompting El Paso's 1982 decision to seek buyers for some of its assets—specifically its company-owned gas reserves. But before any deal was concluded, Burlington Northern, Inc. swallowed the entire company.

El Paso was not alone in its distress. Columbia, Consolidated, and Southern also had big equity stakes in plant facilities, though they were partially protected by minimum-bill tariffs. All three companies continued negotiations with Algeria long after El Paso had deserted the effort. But in late 1982, Consolidated formally abandoned its 50 percent interest in the Maryland facility and applied to FERC for cost recovery that might treat its equity investment more favorably than continued operation of the minimum-bill tariff.

Consolidated and Columbia were both hit in 1982 with a claim by the state of Ohio that the companies were tardy in reacting to Algeria's cutoff in LNG shipments. Instead of invoking the minimum-bill provisions immediately (thereby eliminating the return on equity from the rate formula), the companies spread the full costs over a reduced volume of daily deliveries. Eight months later, when the stored gas was finally exhausted, minimum billing took effect. If the state succeeds, the two transmission companies could face rebate orders of up to $27 million.

By 1982, when the U.S. gas shortage had turned into a gas glut (thereby prompting abandonment of most supplemental gas plans), only one other base-load LNG project in the United States had made it past the promotional and design stages. As of mid-1983, however, the outlook for the Trunkline LNG project was decidedly grim.

In 1982, Algeria delayed commencement of deliveries to Trunkline LNG Company (a subsidiary of Panhandle Eastern Pipe Line Company) pending resolution of a contract dispute which threatened the twenty-year, 165 bcf-per-year supply commitment. An agreement was reached soon after Panhandle petitioned the International Chamber of Commerce for arbitration. Trunkline's regasification terminal at Lake Charles, Louisiana, received its first shipments in September 1982, thereby activating the tariff and ensuring Trunkline of minimum-billing protection.

At a price f.o.b. Algeria of $3.92 per mmbtu, the gas cost $7.16 at the outlet of the regasification terminal—at a time when industrial customers were fleeing to alternative fuels because retail gas prices were approaching $3.50 per mmbtu. The LNG project, therefore, had a powerful effect on gas prices in states connected to the Trunkline and Panhandle systems, most notably Michigan, Ohio, and Pennsylvania. Algerian LNG accounted for a 40 percent rate hike in Trunkline's 1983, first-quarter tariff filing. Not surprisingly, downstream distributors and their congressmen and state utility commissions launched a forceful protest.

At stake was $567 million invested in the regasification terminal, of which about a third (or 14 percent of Panhandle's net worth) was equity. In addition, Panhandle held a 40 percent interest in two LNG tankers that represented close to $200 million. (The other three tankers dedicated to the project were owned by Algeria.)

Although the tariff format approved at the time of certification was designed to ensure recovery of sunk debt and equity irrespective of the fate of the project, the protestors found a legal handle by which to force a review of the merits of the import scheme. Opponents of the project (many of whom had publicly supported Trunkline's plans during the initial FERC certification proceedings) claimed that because the Algerian supply contract had changed since the certificate was issued, regulators had grounds for nullifying the transaction. Some argued that regional consumers would be better off if the supply contract were cancelled so that downstream customers had to absorb only the costs of the regasification facilities rather than $7.16 per mmbtu for gas they did not need.

On the other hand, those who believed that the minimum-bill tariffs should be honored based their arguments on the need to maintain investor confidence in FERC certificates. Specifically, they feared that subsequent attempts by utilities to finance large construction projects would prove difficult if regulators disallowed cost pass-throughs in this instance or otherwise tinkered with the tariff upon which financing had been secured.

Although an administrative law judge ruled against the petitioners in early 1983 (thereby ensuring flow-through of the full LNG charges for the present), the issue remains unsettled until the commission and then (most likely) the courts take action. Soon after the administrative law judge handed down the decision, however, the commission encouraged Panhandle to press Algeria for a price reduction and announced that it would delay its own ruling until Panhandle had a chance to act.

Few industry observers see any hope for baseload LNG-import projects in the foreseeable future. A frustrating record of siting and pricing difficulties is exacerbated by a U.S. supply-and-demand situation in the 1980s that bears little resemblance to what prevailed during the previous decade. Moreover, North America as a whole is far more likely to become a major gas exporter than a gas importer. Although political exigencies sometimes prompt actions that defy logic, the business and financial communities would be hard pressed to support another base-load LNG import scheme while Canada and Mexico struggle to market their shut-in or flared gas (even to the point of considering LNG export projects of their own).

## HIGH-btu COAL GASIFICATION

During the 1970s, interstate transmission companies pursued base-load coal-gasification projects with about as much vigor as they devoted to LNG projects. And like LNG, promotion of coal gasification met with little real success. Not until 1981 did any project reach the ground-breaking stage.

Before we explore the turbulent history of coal-gasification ventures, however, it is important to clear up some confusion in terminology. Coal gas is often referred to simply as *SNG*, short for *synthetic natural gas*—an apt though rather nonsensical phrase. SNG in other contexts means *substitute natural gas*, referring to peak-shaving supplies of energy vapors produced from propane, butane, or naphtha. Despite these ambiguities, we too will use SNG as shorthand for base-load coal gasification.

The coal-gas projects proposed in recent times are technologically and financially different from the coal-gas works that sprung up around the country a hundred years ago. To gain access to the nation's pipeline network, coal gas today must be interchangeable with natural methane. The old kind of manufactured gas, which offered only 300 to 500 btu per cf, would be unacceptable today. Pipeline standards now require gas to contain about 1,000 btu per cf, imposing challenging and costly demands on coal-gasification technologies.

Both the federal government and the transmission industry channeled a lot of money into gasification research during the 1970s. Federal regulators not only allowed regulated pipelines to include small research and development (R&D) projects in their rate bases but encouraged consumer financing of research through rate-base treatment of utility contributions to the Gas Research Institute, created in 1978. But commercial ventures cost about a hundred times more than R&D projects. With price tags in the billion-dollar range, no track record, and with per-unit costs of production exceeding expected market values, financing of commercial-scale *demonstration plants* was far from easy.

Sponsors of one of the earliest proposals spent six fruitless years on project promotion. In 1973, Pacific Lighting Corporation and Texas Eastern Transmission Corporation proposed to build a $1.3 billion plant on Navajo lands in New Mexico. Federal approval of the WESCO plan came two years later, but the sponsors were denied their request for a full *cost-of-service tariff*. Consequent difficulties in putting together the financing, coupled with disagreements with the Tribal Council that owned the coal and the processing site, forced the sponsors to abandon their plans in 1979.

Likewise, in March 1982 Panhandle Eastern Pipe Line Company scrapped its ten-year-old proposal for a gasification plant in Wyoming. Citing inflation and high interest rates that boosted the WyCoal Gas project cost to $3.5 billion, Panhandle figured that estimated product prices equivalent to oil at $100 per barrel made the project too risky.

## The Great Plains Project

The only proposal to reach the construction stage was the Great Plains project, whose principal sponsor was American Natural Resources Company (parent of Michigan-Wisconsin Pipeline Company). Groundbreaking for the North Dakota facility occurred in 1981, with completion scheduled for 1985.

The Great Plains SNG concept encountered at least as many frustrations as did El Paso's LNG venture. It started out as a 250 mmcf-per-day facility but was pared down to 125 mmcf per day. For the first few years it was solely sponsored by American Natural Resources Company, operating as ANG Coal Gasification Company. Ownership was broadened in early 1978 to include three other interstate transmission companies: Tenneco (parent of Tennessee Gas Transmission Company), Transcontinental Gas Pipe Line Corporation (Transco), and Columbia Gas Transmission Corp. The partnership adopted the name Great Plains Gasification Association.

Although the plant site in Mercer County, North Dakota, was never subject to dispute, it took several years to determine the most appropriate means for moving the gas to market. American Natural preferred to build a 365-mile spur to its subsidiary pipeline, Great Lakes Transmission Company. Federal regulators, however, decided to leave the question open pending forthcoming decisions about the Northern Border pipeline. As the eastern leg of the planned Alaska Natural Gas Transportation System, Northern Border's sponsors had designated a corridor that came within twenty miles of the gasification site. In 1982, when Northern Border began carrying Canadian gas, that system became the anticipated carrier for Great Plains SNG.

The major dilemmas confronting the Great Plains project centered on tariff and financing questions. From the very beginning, the sponsors made it clear that the nation's first commercial-scale coal gasification plant would require federal support in the form of loan guarantees. But Congress was slow to act, and the partnership approached federal regulators in 1976 with a tariff concept that promised investors financial security by transferring much of the risk downstream.

FPC staff vigorously objected to some of the proposed financing and tariff provisions, and late in 1976, American Natural requested that formal hearings be postponed. The company anticipated congressional passage of loan guarantees that would make the most controversial tariff elements unnecessary. The legislation failed by a hair, however, and American Natural was forced to renew its administrative pursuits.

Determined to finance the plant without federal loan guarantees, American Natural broadened its ranks to include the three interstate transmission companies previously mentioned. The new partners strengthened the project's equity base and would be able to dilute the inflationary effects of high-cost gas by channelling the product into several pipeline systems. Nevertheless, this revised approach still depended on regulatory acceptance of some novel tariff provisions.

## Regulatory Troubles for Great Plains

Historically, transmission companies have operated under *fixed-price tariffs* by which a sales price was established when the FPC approved pipeline construction. But Great Plains proposed a special form of *cost-of-service tariff* that gave the owners license to adjust their prices without a rate hearing if the actual costs exceeded preconstruction estimates.

Although gas supplies have always been treated on a *rolled-in* basis by the purchasing pipeline companies, FPC staff advocated that the SNG be handled incrementally. Federal policymakers had begun to think it prudent to load the new, high-cost projects disproportionately onto industrial customers, who were indeed *marginal users* of gas. The transmission industry challenged that idea for several reasons.

Industrial customers, they argued, played a vital role as off-season purchasers. Absent the *interruptible sales*, year-round residential and commercial customers with low summer takes would face higher per-unit transportation charges. (See Chapter 7 for a detailed explanation of the role of interruptible purchases.) Because they were interruptible, and therefore able to switch to other fuels, industrial users placed the lowest value on gas. Great Plains rightly feared that if forced to enter into specific contracts with industrial customers for sale of the coal gas and at a price undiluted by other gas supplies, market demand might not support the project.

Whatever the philosophical differences between sponsors and FPC staff, practical considerations finally held sway. The tariff structure attendant to incremental pricing was incompatible with federal *curtailment policies*. On the one hand, incremental pricing meant that industrial customers would specifically agree to purchase SNG on a year-round basis. But if that were the case, gas companies would no longer be able to curtail those same *low-priority users* during seasonal supply shortages.

Then too, incremental pricing would be easy to arrange only for industrial customers who bought gas directly from the interstate pipeline. But how did one reach an industrial user who happened to buy gas from a local distributor served by the same pipeline? And if one could not, was it fair to impose the added costs of incremental pricing only on the former?

Another controversy concerned the rate of return on equity investment. If commercial-scale coal gasification was so risky that it demanded federal loan guarantees or consumer support, what was a "just and reasonable" profit? On the other hand, if a federal guarantee was forthcoming, would not the project be even less risky than conventional investments? Great Plains applied for a 15 percent return. FPC staff advocated a lower rate.

Perhaps the most unpalatable provision of the *pro-forma tariff* submitted by Great Plains was the "all-events" transfer of risk downstream. In addition to the *minimum-billing* concept (first applied by the El Paso LNG venture), which ensured recovery of capital if plant operations were jeopardized, the Great Plains tariff dealt specifically with the risk of *noncompletion*. The revised tariff filed in 1978 provided that in the event the project were abandoned during the construction phase, all debt money (and corresponding interest) invested in the project would be paid back by customers of the sponsoring pipeline companies over a five-year period. Sixty percent of the contributed equity capital would be recovered in the same fashion.

The downstream transfer of risks under all events, including non-completion, was loosely called a *consumer guarantee*. Ultimately, however, consumers do not and can not "guarantee" anything. If a supplemental venture, like an SNG plant, should fail, and if the price of gas were, as a result, to rise beyond its market value to industrial customers, consumers could switch to other fuels. Their distributors are, however, far less flexible. If customers begin to reject gas in a big way, the system's fixed costs must be spread over fewer units of gas sold, which in turn increases the unit price and discourages even more customers, and so on.

To ensure that noncompletion billing would indeed take effect, the sponsors proposed a novel approach to *AFUDC (allowance for funds used during construction)*. Instead of capitalizing the interest on debt and return on equity that accrued during the construction period, the consortium proposed a *surcharge*. Customers of the sponsoring pipelines would pay the surcharge on a current basis during construction, and if abandonment ensued, Great Plains could simply inflate the surcharge rather than having to subject their project to a new tariff hearing.

The squabble between commission staff and the Great Plains consortium was not, however, limited to tariff provisions. A dispute over government access to financial documents prompted the staff to request outright dismissal of the Great Plains application. The administrative law judge ruled against the staff on this matter, but nevertheless recommended that the commission reject the project as it was then structured. He reasoned that a demonstration facility would benefit the entire nation and that it was therefore unfair that only a portion of the nation's gas users should bear the brunt of project risks. Instead, he argued that the nation as a whole should support the project through a Treasury loan guarantee.

This decision came as a tremendous blow to the consortium. Engineering and design came to a halt, prompting the Department of Energy to make a direct $3 million contribution for its resumption. In defending this action, Deputy Undersecretary Hanfling told a congressional committee that $3 million was a pittance compared to the capital that the transmission industry had spent on coal gasification development and promotion during the past half dozen years. Company investment of more than $100 million was at stake.

Five months later, in November 1979, the commission (now reconstituted as the Federal Energy Regulatory Commission, or FERC) overturned the judge's recommendations and approved the Great Plains project. Basically, it granted the consortium nearly all it was seeking. FERC did, however, limit the return on equity to 13 percent, subject to revision once operations began. The approval was also conditioned upon a commitment by Great Plains to rejuvenate its quest for a federal loan guarantee. The commission felt that federal and consumer guarantees together would distribute the risks equitably among the sponsors, the consumers, and the nation as a whole.

### The Quest for a Federal Loan Guarantee

The situation looked promising for Great Plains at this point. The Department of Energy and many members of Congress supported a bill that would authorize federal loan guarantees for synthetic fuels projects, including coal gasification. In June 1980, the *Energy Security Act* became law (S. 932, P.L. 96-294). It established the U.S. Synthetic Fuels Corporation, with authority to grant up to $88 billion in federal financial assistance, including loan guarantees. To ensure progress before the organization was in place, amendments to the Defense Production Act of 1950 awarded the Department of Energy immediate power to distribute up to $3 billion.

Meanwhile, four parties had filed an appeal to nullify FERC's certification of the Great Plains project. Public utility commissions of New York, Michigan, and Ohio, along with General Motors Company, argued against the consumer guarantee. In December 1980, the U.S. Court of Appeals for the District of Columbia Circuit ruled in favor of the plaintiffs on several grounds.

As in many legal disputes, it is often obscure technicalities that give an aggrieved party power to negotiate changes in substantive matters. The Great Plains decision was a case-in-point. Because SNG is not in fact "natural" gas, the judge found that the Natural Gas Act of 1938, under which the commission operates, did not grant FERC any power to certify the synthetic gas plant nor to authorize interstate pipelines to pass SNG costs downstream through their Purchased Gas Adjustment clauses.

This technical ruling was reinforced by a policy ruling that, as the administrative law judge found three years earlier, found the scale of consumer risk-bearing unwarranted. Moreover, in light of the now-imminent Treasury guarantee pursuant to the new law, the judge held that the consumer guarantee was unnecessary.

After several unsuccessful rounds of negotiations, the plaintiffs, the consortium, and FERC reached agreement in April 1981. Basically, the cost-of-service tariff was converted into a fixed-price tariff (beginning at $6.75 per mcf) that would move with changes in fuel oil prices. Like all other tariffs, it would be open to revision via the standard regulatory process. The consumer noncompletion guarantee, however, was eliminated.

The size of the prospective loan guarantee swelled within the course of a year. When the Energy Security Act was passed, the Department of Energy announced that it would seek presidential concurrence, pursuant to the act, of a $250 million guarantee for Great Plains. Five months later, in November of 1980, that figure rose to $1.5 billion, then to $1.8 billion in February of 1981. Finally in August, President Reagan approved a $2.02 billion guarantee to cover project debt, a portion of the equity, and possible cost overruns.

Subsequently, American Natural Resources and Transco both diluted their equity positions by selling to Pacific Lighting Corporation stock totalling 10 percent of the venture. With its pipeline system limited to California, Pacific Lighting is the only equity participant that is not committed to purchasing the manufactured gas.

With construction underway, overruns do not appear likely. In fact, accounts in mid-1983 showed that the project might be completed under budget. Nevertheless, the recent upset in the decade-long trend of rising oil prices, combined with rapid depletion of the "cushion" of cheap gas, quickened old fears about gas marketability.

In March 1983, the company reported that it had revised its cash-flow analysis based on the new oil-price outlook. Replacing a projection of $1.2 billion in earnings during the first decade of operations, the partners now envisioned a $770 million loss. Even that figure, however, may be optimistic—for it is grounded on the assumption that oil prices will continue to increase throughout those ten years at an average annual rate that is 5 percent above the national inflation rate.

All in all, while the Great Plains Project is likely to be a technical success and a sterling example of on-budget, on-time construction management, the economic misjudgments of its corporate leaders are likely to overwhelm the talents of the engineers.

## ARCTIC GAS PIPELINES

Of all the supplemental projects proposed in the 1970s, the Alaska Highway gas pipeline was the most ambitious. At a projected cost of $25 billion in 1982 dollars—and $40 billion in "as-spent" dollars (see Table 4–2)—an Alaska gas pipeline would require a far greater influx of capital over its five years of construction than the cumulative plant investments of the entire interstate gas industry. (The American Gas Association, in a December 1982 report, pegged the latter at about $25 billion in as-spent dollars throughout the half-century of construction activity.)

Sponsors touted the project as potentially the largest private investment in the history of the world, surpassing the $9 billion that had gone into the Trans-Alaska (oil) Pipeline System (TAPS) during the mid 1970s. But this claim to fame was also the project's downfall. In April 1982, four-and-a-half years after the sponsors of the Alaska Natural Gas Transportation System (ANGTS) received a "conditional" certificate to go ahead with the project (and fully eight years after federal regulators had commenced hearings on routing and

**Table 4–2.**  Estimated Capital Costs of the Alaska Natural Gas Transportation System.

| | Construction Costs in 1982 Dollars | | Capital Costs in Current-Year Dollars with Interest[a] |
| | Base Estimates | With Contingencies | |
| --- | --- | --- | --- |
| | *(billions)* | | |
| Alaska pipeline | $ 8.6 | $ 9.9[b] | $17.2 |
| Gas conditioning plant | 3.4 | 4.1[c] | 7.0 |
| Canadian pipeline | 5.9 | 7.4[c] | 12.1 |
| Lower 48 states pipeline | 3.4 | 3.4 | 4.6 |
| Total | $21.3 | $24.8 | $40.9 |

Sources: For 1982 estimates: Participant companies; Alaska pipeline costs taken from FERC cost certification order of February 18, 1983, plus some costs deferred. U.S Comptroller General, General Accounting Office, *Issues Facing the Future Use of Alaskan North Slope Natural Gas.* (Washington, D.C.: U.S. Government Printing Office, May 1983).

[a] Compiled by GAO, using interest and inflation rates found in appendix III. Estimates were factored from construction cost with contingencies.

[b] Assumes FERC-approved, 12 percent contingency to cover normal uncertainties associated with estimating costs of materials, labor, equipment, and so forth, and unexpected or unlikely uncertainties such as earthquakes or sabotage.

[c] Assumes sponsors' 20 percent contingency.

sponsorship), ANGTS was put on indefinite hold. The reason? Failure to secure the requisite financing from "private" sources outside of the federal treasury.

The failure of the ANGTS project (at least for the foreseeable future) was one of the biggest disappointments for the interstate gas industry and for American advocates of energy independence. Prudhoe Bay reserves in northern Alaska totalled 26 tcf, or about 13 percent of remaining U.S. proved reserves. Then too, Alaska accounted for more than half of the nation's Outer Continental Shelf (OCS), and official estimates of gas potential in OCS basins ranked the area seaward of the onshore wells at Prudhoe Bay (the Beaufort Sea) as the most promising of all.

Soon after the project was officially put on hold, the initial disappointment of ANGTS supporters was surely replaced by a sense of relief. After all, the few supplemental-gas projects that were financed successfully proved to be losers in the marketplace. The financial disaster visited upon El Paso Company and the other sponsors of the Algeria I LNG project was already history, while there was no reason to expect fate to be any gentler with the embattled Trunkline LNG venture nor with the Great Plains coal gasification plant, then under construction. What is more, by the end of 1982, the United States was in the midst of a gas-consumer "revolt" unlike anything yet experienced. (Refer to Chapter 7.) Pipeline companies were finding it impossible to sell all their available gas at prices only a third as high as what they had planned to charge for Alaska gas only a few years later.

Actually, by the fall of 1982, about 1,500 miles of the 4,800-mile binational system were in place and carrying gas. (See Map 4-1.) These southern portions of ANGTS were, however, conventional with respect to scale and climatic and physiographic conditions. And the entire U.S. and Canadian effort had cost "only" about $1.6 billion. Then too, these *prebuild* sections of the Alaska pipeline were designed to carry surplus volumes of Alberta gas into U.S. markets, pending completion of facilities across Alaska and through the Yukon Territory. The depreciation schedules in both countries, therefore, promised investment recovery roughly matching the seven-year period for which the Canadian gas-export licenses had been assured. *Ship-or-pay* commitments by the U.S. pipeline purchasers (who intended to transport Canadian gas through the prebuild system on a *contract* basis) also supported the successful financing.

**Map 4-1.**   Alaska Natural Gas Transportation System.

Even the ANGTS prebuild, though completed on time and on budget, was not the glowing success that its sponsors had anticipated. Project start-up coincided with a growing glut of gas in the United States, and within six months, the Eastern leg of the prebuild (officially known as the Northern Border Pipe Line) was transmitting only 30 percent of the contracted volumes. Canadian interests were hurt not only by reduced throughput in the 530 miles of new pipeline within Canada but also by the fact that TransCanada Pipeline Company was, itself, the largest of the five corporate shareholders of the 820-mile Northern Border system in the United States. And because of Canada's "netback" pricing arrangement, Alberta gas producers felt the pinch of both dwindling sales and the high costs of transmission capacity. In May 1983, Premier Peter Lougheed claimed that shipping gas to the United States through the prebuild pipelines cost $1.00 to $1.50 more than it would have cost if existing facilities had been looped instead—and that this amount came right out of the Alberta producers' pockets (and the province's tax and royalty income).

Although the ANGTS project is still dutifully supported by both Canadian and U.S. governments, and although only two of its pipeline cosponsors had officially withdrawn by mid-1983, it has taken a tumble from the exalted status it held in the late 1970s. At one time, the ANGTS project boasted fourteen U.S. cosponsors, including eleven of the biggest interstate pipelines and the three oil companies with the biggest stake in Prudhoe Bay hydrocarbons. (See Table 4-3.)

**Table 4-3.**    ANGTS Sponsors.

---

*U.S. Pipeline Companies:*

Northwest Pipeline Co.
InterNorth Inc. (Northern Natural)
Panhandle Eastern Pipe Line Co.
United Gas Pipeline Co.
Pacific Gas & Electric Co.
Pacific Lighting Corp.
Texas Eastern Transmission Corp.
Columbia Gas Transmission Co.
Texas Gas Transmission Corp. (withdrawn)
Michigan-Wisconsin Pipe Line Co. (withdrawn)

*U.S. Gas Producers:*

Exxon
Atlantic Richfield Co. (ARCO)
Standard Oil of Ohio (SOHIO)

*Canadian Pipeline Companies:*

NOVA Corp. (formerly Alberta Gas Trunkline Co.)
Westcoast Transmission Co.
TransCanada Pipeline Co.
Alberta Natural Gas Co. (Canadian affiliate of Pacific Gas & Electric)

---

Virtually the entire pipeline industry of western Canada was represented too, with four transmission companies sharing equity positions for those portions of the system that would traverse Canadian soil. Congress passed special legislation in behalf of ANGTS on four occasions (see Table 4-4), and the United States entered into a formal compact with Canada at the outset. With so much going for it, what indeed went wrong?

### Physical and Financial Risks

The quest for a gas pipeline from Arctic Alaska began soon after Atlantic Richfield Company stumbled upon the biggest hydrocarbon

**Table 4-4.**   Legislation Supporting ANGTS.

---

*1976: The Alaska Natural Gas Transportation Act of 1976*

ANGTA limited administrative and judicial review of licensing and permit procedures; it provided for presidential selection of a route and sponsorship and for congressional approval thereof.

*1977: Joint Resolution concurring with the President's Decision*

In May, the Federal Power Commission recommended to the president selection of an overland route through Canada, with final designation of either Alcan or Arctic Gas awaiting Canadian recommendation. Canada subsequently endorsed Alcan, and President Carter likewise selected Alcan in September. Congress adopted the president's decision in November.

*1978: The Natural Gas Policy Act, Section 109*

Section 109 established a wellhead ceiling price for Prudhoe Bay gas and assured ANGTS sponsors that the usual "rolled-in pricing" would apply.

*1980: Joint Resolution reiterating U.S. support for ANGTS*

This resolution passed in July in order to quell fears in Canada that if the "prebuild" took place and Canadian gas exports therefore expanded, completion of the northern segments of ANGTS might be indefinitely delayed.

*1981: Joint Resolution approving the "waiver package"*[a]

This resolution waived certain prohibitions that had been established by the 1977 presidential decision and accompanying congressional resolution, including prohibition of producer equity participation and limitations on consumer risk-bearing in ANGTS tariff.

---

[a]Legislation was introduced in 1983 that would have caused the waiver law to expire at year-end unless FERC had issued a final certificate to ANGTS.

discovery then known in North America. The construction priority following the 1968 Prudhoe Bay find was, of course, a system for oil transportation. The 800-mile TAPS oil pipeline, spanning Alaska from the Arctic ocean to the southern Gulf (see Map 4-2), "broke ground" in many ways for a counterpart system to deliver gas. Congress had settled the long-overlooked land claims of the Indian and Eskimo peoples native to the area in 1971, and two years later it resolved the environmental impasse over pipeline construction. Then, too, TAPS proved that the physical and engineering challenges posed by one of the world's most remote regions and harshest climates were not insurmountable.

**Map 4-2.**    The Three Competing Proposals.

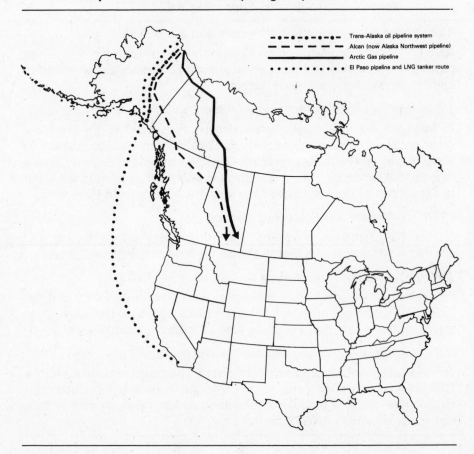

Trans-Alaska oil pipeline system
Alcan (now Alaska Northwest pipeline)
Arctic Gas pipeline
El Paso pipeline and LNG tanker route

The legacy of the TAPS construction was not, however, altogether positive. By mid-1977, when the president and Congress selected the configuration and industry sponsors for a gas delivery system, TAPS construction was complete—but at a cost that was ten times higher than initial estimates. Had it not been for the fivefold rise in world oil prices during the construction period, the Alaska oil pipeline would have been a financial disaster.

Federal officials saw a gas pipeline as vulnerable to similar cost overruns. Moreover, favorable market forecasts for Alaska gas were based in large part upon two assumptions: that the upward trend of world oil prices was permanent and that the regulatory edifice would continue to suppress domestic wellhead prices for gas, thereby ensuring a long-lived "cushion" of low-cost gas to offset the above-market costs of supplemental supplies. While successful marketing of Alaska gas might have seemed a reasonable gamble, it should have been clear at the outset that investors and lenders would be reluctant to bet tens of billions of dollars—putting their very futures at the mercy of these assumptions. Quite simply, an Alaska gas transportation system was too unconventional a project for conventional approaches to financing.

During the three years of federal hearings that led to the 1977 issuance of a conditional certificate, these financial questions took a backseat to the choice of route. Indeed, of the three competing proposals (see Map 4-2), the group of companies advocating construction of a pipeline along the Alaska Highway were financially the least capable of carrying out their plan. Nevertheless, Congress and the president selected the Alcan proposal (reorganized soon after as Alaskan Northwest Pipeline Company) and awarded the certificate for an Alaska Natural Gas Transportation System to Northwest Pipeline Company, the single U.S. firm participating in the binational consortium.

Northwest was not only a small company compared to the industry giants supporting the two other applications, but it was a last-minute entrant. Being small and coming along late, however, did have its advantages. Northwest and its two Canadian pipeline cosponsors (Westcoast Transmission Company and Alberta Gas Trunk Line Company, now NOVA) were able to put together a superior proposal, based upon careful observation of the criticisms leveled against Arctic Gas and El Paso. When physical or political flaws in its own proposal became apparent, the Alcan sponsors not only were willing to alter their plans but were free of the bureaucratic obstacles to decisionmaking that plagued the multimember Arctic Gas group.

Within a few years of its congressional victory, however, the Alaskan Northwest partnership had lost much of the flexibility that had made it a winner. Some of the rigidity came from the inevitable growth in membership; at its peak the consortium included eleven U.S. pipeline companies, four Canadian pipeline companies, and three major oil interests. Although El Paso never sought an ownership share, most of the gas pipelines and all of the U.S. oil companies that had been the driving forces behind Arctic Gas did make the switch.

Remarkably, however, John McMillian, chairman of the board of Northwest Pipeline's parent company and unquestionably the most colorful (and perhaps successful) entrepreneur within an otherwise dreary industry, retained both his power and his public prominence. John McMillian and the Alaska gas pipeline were virtually synonymous. Unfortunately, this link meant that while the growing ANGTS consortium might have found it possible to retract on earlier promises made by the fledgling and eager-to-please Alcan, its chief executive officer and spokesman could not. What were these promises and why were they so debilitating?

## Promises to Keep

In the mid-1970s, the notion that a transportation system for Alaska natural gas could meet conventional market and cost-benefit tests was not terribly controversial. But it was also generally believed that financing such a system would require extraordinary measures to shift construction and operating risks to consumers, the federal government, or both. Indeed, by 1977 all three of the supplemental gas projects that would eventually come to fruition (Algeria I LNG, Trunkline LNG, and Great Plains Coal Gasification) had made no pretense about their needs for consumer and/or governmental risk sharing.

The two original applicants for an Alaska gas pipeline certificate, Arctic Gas and El Paso, likewise had been explicit about the need for special assistance. In addition to rolled-in pricing, they maintained that such a project would require both a consumer guarantee in the form of an "all-events, full cost-of-service tariff" and a federal promise to make good on scheduled debt payments in the event that construction was abandoned or the completed project experienced a shortfall in revenues.

Alcan, however, asked neither for an all-events tariff nor for federal loan guarantees. Northwest's contention that such assistance was unnecessary was plausible because the Alcan plan was less risky than its rivals. In its original form at least (which proposed looping existing Canadian and American facilities wherever possible rather than installing a completely new system), the Alaska Highway project required the smallest capital outlay of the three proposals. Its frontier sections would be built along an existing pipeline (TAPS) and existing highway (Alcan), using conventional technology and conventional construction methods. Northwest's Canadian collaborators (known as the Foothills group) had experience in building gas pipelines through rugged mountains, muskeg, and permafrost—almost always completing them on time and within budget. And the project was relatively immune from controversy over Native land claims, environmental impacts, or safety. Moreover, with the addition of a small-diameter lateral from the Mackenzie Delta southward along the Dempster Highway, the project rivaled Arctic Gas with respect to potential access to frontier reserves of Canadian gas.

Not only was the Alcan scheme less risky in a physical sense, but Northwest made a valiant effort to devise approaches for easing the financing risks as well. Almost from the start, Northwest proposed an overcollection of construction capital (an *overrun pool*) to reduce the chances that in the event of an overrun, subsequent financing obstacles would threaten project completion. Northwest also readily accepted federal imposition of an *incentive rate of return (IROR)*, which was part of the certification package approved by Congress in 1977. (The IROR concept, the brainchild of Professor Walter Mead, was intended to stifle any tendency toward goldplating or careless project management by awarding a progressively higher rate-of-return if the system was completed on or under budget. Its perverse effects, however, may have outdistanced its potential cost savings: The IROR may have simply created an incentive to inflate the blueprints rather than actual expenditures.) Later, Northwest conceived of the prebuild—an ingenious plan to preserve project momentum, to ease the pressure on Wall Street by segmenting capital acquisition, and to ensure sponsors' recovery of at least some of the hundreds of millions of dollars already spent on project planning and promotion. (In May 1983, the General Accounting Office pegged the latter at $750 million for U.S. sponsors and $250 million for Canadian sponsors).

In essence, the Alcan group distinguished itself from the other two applicants by its willingness to explore ways for reducing the uncertainties and risks, rather than simply loading those risks onto consumers and the government. Unfortunately, that willingness to strive for nonconsumer and nongovernment risk taking became a promise of achievement. In his 1977 *Decision and Report to Congress* (Executive Office of the President 1977), President Carter stated:

> The Alcan sponsors and their financial advisors have stated that the Alcan project can be privately financed. . . . Novel regulatory schemes to shift this project's risks from the private sector to consumers are found to be neither necessary nor desirable. Federal financing assistance is also found to be neither necessary nor desirable, and any such approach is herewith explicitly rejected.

The president's forthright treatment of the financing issue was not unexpected. The action was presaged by a congressional directive (through the Alaska Natural Gas Transportation Act of 1976) for the president to not only recommend to Congress his choice of route and sponsorship, but to submit:

> . . . a financial analysis of the transportation system designated for approval. Unless the President finds and states in his report . . . that he reasonably anticipates that the system he selects can be privately financed, constructed and operated, his report shall also be accompanied by his recommendation concerning the use of existing Federal financing authority or the need for new Federal financing authority.

The president's choice was affirmed by Congress and his *Decision* incorporated by reference. Congressional leaders were not entirely convinced, however, of the prospects for a successful private financing of this massive project, and they made a special effort to record their concerns and intentions. The report of the Senate Energy Committee approving the presidential decision is worth quoting at length:

> While the Committee has reservations about the ability of the Alcan project sponsors to secure the necessary private financing, we are recommending approval of the President's *Decision* based upon the unqualified assertions made by the administration and Alcan officials.

It is essential for the project's sponsors to proceed with their financing arrangements as promptly as possible. The State of Alaska, the producers, and most of all the project sponsors should bear in mind that the door to the Federal Treasury has not been left open to them. We have taken the administration's and the sponsors' assurances at face value and are placing our reliance upon them.

The Committee cautions the Administration and the sponsors against taking a back door approach to Federal financing. We are, of course, aware of the possibility that the Federal Energy Regulatory Commission may be tempted to devise a new type of tariff, or a special type of well-head price policy, that would in essence be a "back door" or indirect approach with the same practical effect as direct Federal participation in project financing. We intend to monitor the project's progress closely and caution that financial "gimmicks" involving consumer risk-taking via the Federal Treasury or via special tariffs will not be tolerated by the Congress.

The proposition that there would be no direct federal assistance in the form of loan guarantees or gas-purchase guarantees thus began as a hope and objective of federal policy, emerged as a promise by the Alcan sponsors and President Carter, and was transformed to a ritual oath by Congress.


### Too Little Too Late

The first crack in the proscription of consumer or government support was achieved in the Natural Gas Policy Act of 1978, where ANGTS was assured an ability to sell gas on a "rolled-in" basis. However, the Congressional Conference Report stated, "Rolled-in pricing is the only Federal subsidy, of any type, direct or indirect, to be provided for the pipeline." The following year, Northwest's management persuaded the state of Alaska (a one-eighth royalty owner of Prudhoe Bay gas) to set up machinery to issue tax-exempt revenue bonds for the Alaska segment of ANGTS, but it never actively pushed for the requisite amendments to the Internal Revenue Code.

It was more than two years later, in the 1981 "waiver package," that the ANGTS sponsors were to ask for congressional concurrence in regulatory tactics that were unabashedly consumer guarantees. Late that year, Congress authorized a minimum-bill tariff similar to what was already deployed by the two big LNG ventures. Congress also waived the legal barriers that had prevented gas consumers from bearing the risk of project noncompletion; the U.S. and Canadian tariffs could now incorporate a *pre-billing* clause to ensure that revenues would be forthcoming even if gas was not. Finally, the 1981 congressional resolution waived the 1977 congressional mandate that prohibited equity participation by the gas producers.

It was thus not until four years after the sponsors had received a conditional certificate that ANGTS was granted essentially the same level of consumer support that its two LNG predecessors had been awarded in the mid-1970s. And as of mid-1983, ANGTS had never sought the kind of federal loan support that investors in the much smaller Great Plains coal-gasification plant had consistently maintained as essential to its own financing (and which they actually received only a few months before the ANGTS waiver hearings).

Thus the promises that the sponsors of ANGTS made in 1977 came back to haunt them—and still do. Perhaps the greatest damage caused by these promises has been the delay the project suffered while the sponsors tried in vain to achieve what most knowledgeable observers expected right from the start was an impossible goal. Conditions favorable to any kind of private financing for an Alaska gas pipeline—or to any kind of federal aid that might have been sufficient to permit a partial private financing—have probably come and gone.

Unfortunately for the ANGTS sponsors, the delay in project startup coincided with several national and international events over which they had no control. The three most crucial developments, unanticipated by the sponsors and by most governmental bodies concerned with ANGTS, have been (1) a fundamental revolution in the structure and behavior of natural-gas markets in North America, (2) an interruption (and, possibly, the end) of the rise in world oil prices that began in 1973, and (3) the financial collapse of the only three "successful" supplemental gas projects and the consequent death of consumer guarantees as a credible financing tool. (The first two matters are explored in detail in Chapter 7.)

## SUPPLEMENTAL GAS: AN EPITAPH

The supplemental-gas projects of the 1970s began as creative and seemingly intelligent responses to the frightening shortage of domestic natural gas. Because producers were restricted from raising their prices for conventional gas, regulation thereby prevented either a supply or an end-use pricing response to the market imbalance. Transmission companies, however, were allowed (and even encouraged) to invest in supplemental projects that could do both. Foreign LNG, coal gasification, and Arctic pipelines all promised to bolster supplies, and they all promised high costs which, on a rolled-in basis, would increase systemwide gas prices.

What made the supplemental-gas fad so unhealthy, however, was the fact that while companies were investing in capital-intensive projects that could provide gas at perhaps five or ten times the going rates, a huge volume of conventional gas at only two or three times current prices was out of reach. Despite the fanfare, supplemental projects were doomed from the start because their viability rested upon regulatory quirks rather than a fundamental economic logic. Only for so long as the regulatory edifice maintained its essential character would these multibillion dollar projects make sense. With passage of the Natural Gas Policy Act of 1978, however, a fatal crack in the regulatory structure appeared and the visions that had put twinkles in the eyes of gas utility executives now brought only tears.

Regulation was thus responsible for the birth of the supplemental gas projects, and regulatory changes prompted their demise. The next chapter takes a closer look at the evolution of gas-industry regulation in the United States.

## REFERENCES

Executive Office of the President. *Decision and Report to Congress on the Alaska Natural Gas Transportation System.* Washington, D.C.: U.S. Government Printing Office, September 1977.

# 5 THE GROWTH OF GOVERNMENT INVOLVEMENT
# The History of Regulation in the Gas Industry

Utility-style regulation of the gas industry reached a crescendo in the 1970s. In a desperate attempt to deal with a supply shortage that had afflicted the nation for a half dozen years, Congress passed the Natural Gas Policy Act (NGPA) in 1978. The NGPA instituted a program for phased and partial deregulation of wellhead gas prices. Congress, however, chose to achieve those ends by extending both the reach and the complexity of federal intervention in gas markets. NGPA-style "deregulation" added whole new layers of federal controls, which in turn spawned a new generation of problems and inefficiencies.

Federal regulation of the gas industry is coming to look more and more like a tarbaby. The ink barely dries on a new law or a new rule when it becomes obvious that the intended solution gives rise to altogether new headaches. Who would have guessed that by eliminating the gas shortages and curtailments of the 1970s, the NGPA would give rise to an equally troublesome gas glut and the take-or-pay crunch in the early 1980s? The tables turned on the inter-/intrastate schism too, with interstate pipelines bidding gas away from Texas and Louisiana carriers.

In blatant violation of the spirit behind NGPA's companion legislation, the Power Plant and Industrial Fuels Use Act, gas distributors and transmission companies in 1982 began a mad scramble to keep their low-priority industrial customers happily burning gas. Some companies adopted *value-of-service* rate structures that parodied the incremental pricing provisions of the NGPA. And state commissions fresh from 1970s battles to impose *lifeline rates* on reluctant utilities found they had little choice but to acquiesce to rate strategies that loaded rising gas prices mainly on homeowners.

The chaos in the industry today and the alarming failure of the NGPA provokes either timidity or brashness among those who are in a position to act. Neither approach is responsible. What is needed today is a broader perspective—a sense of history that is not blinded by the immediacy and the drama of the OPEC era.

Only by probing how and why we got into this predicament can our leaders hope to find a way out. An examination of past events and regulatory debates reveals that today's questions are far from unique. Common-carrier status for interstate gas pipelines, for example, was widely discussed prior to passage of the Natural Gas Act in 1938. In the early 1940s, Congress and the Federal Power Commission considered whether and how to implement policies intended to ensure that gas would not be "wasted" in industrial or utility boilers that had traditionally consumed coal. Ever since the 1938 act, regulators and legislators have pondered the proper extent and manner of federal intervention in field transactions, cycling through eras of relaxed and heightened oversight of wellhead pricing and other terms of sale.

Finally, looking beyond particular issues, there emerges a pattern and (in hindsight) a logical evolution of the forces that from time to time prompted changes in federal regulation. From this historical vantage point, we can detect a glimmer of hope (and perhaps even opportunities for progress) in events that would otherwise bring only confusion and despair.

## THE ADVENT OF REGULATION: FRANCHISING LOCAL DISTRIBUTION COMPANIES

As discussed in Chapter 2, government involvement in the affairs of the gas industry began in the mid-1800s, when state and municipal legislative bodies issued charters and franchises for urban coal-gasification plants. Many of these early gas companies chose the corporate form of business organization, in which plant owners were personally immune from financial obligations incurred by the company. Because the country as a whole was wary of this new approach to industrial organization, for many years corporate status depended on state *charters* issued on a case-by-case basis.

Coal-gas companies also lobbied local authorities to gain access to city streets for their distribution mains. Sometimes municipal involvement was important from a market standpoint too. A fledgling company often needed assurance from the outset that somebody would purchase a substantial volume of gas, and municipal street lighting fit the bill. Moreover, until a company was well established, it was vulnerable to cut-throat competition; firms therefore sought special provisions in their charters, rights-of-way ordinances, and municipal supply contracts that granted exclusive *franchises* for a certain number of years.

Governments were willing to grant franchises, thereby restraining competition, because the character of the physical plant made gas distribution a *natural monopoly*. A single company could supply gas at a lower cost than two companies with duplicate facilities and overlapping markets. The key here is "could." A single company *could* provide the gas and especially the transportation service at a lower cost than multiple firms, but in the absence of competition it probably *would* not.

The solution? Gas distribution was declared a business "affected with a public interest." Like other *public utilities*, gas distribution is now subject to governmental oversight of rates.

## STATE REGULATION OF INTRASTATE TRANSMISSION

Local rate regulation was effective as long as manufactured gas was the commodity because it was produced from coal or oil in the same town and usually by the same company that distributed it. But when natural gas from outside the city limits began to enter the market, municipal governments were no longer able to control gas prices. The states stepped in to fill this regulatory gap. State legislatures created public utility commissions and public service commissions or extended the reach of existing trade commissions into the gas transmission business. New York and Wisconsin led the nation, instituting commissions in 1907.

As technology improved and pipelines were able to carry gas several hundred miles or more, a new regulatory gap appeared that even the states could not touch. Between 1911 and 1928, Oklahoma, West Virginia, and Missouri tried to assert jurisdiction over interstate gas transactions but were thwarted each time by federal judges who ruled that state regulation of long-distance gas transmission violated the interstate commerce clause of the U.S. Constitution.

Later, with passage of the Hinshaw Amendment to the Natural Gas Act, Congress exempted from federal jurisdiction pipelines that carried gas in interstate commerce but whose facilities themselves did not reach beyond a single state. Pacific Gas and Electric Company and Pacific Lighting Corporation (both entirely within California) are good examples of *Hinshaw pipelines*. They carry gas delivered to the California/Nevada border by two interstate pipeline companies: El Paso Natural Gas and Transwestern.

## FEDERAL REGULATION OF INTERSTATE TRANSMISSION

Most oil pipelines evolved as *common carriers* or *contract carriers* of crude oil and refined products. Unlike gas transmission companies, which operated almost exclusively on a *private carrier* basis, owners of oil pipelines did not hold title to the goods handled. The courts played a role in limiting the incentive for pipeline ownership of oil shipments when, acting upon the complaints of aggrieved shippers, jurists repeatedly ruled that owners of oil pipelines must not discriminate in favor of shipments by affiliated companies.

In 1906, Congress passed the Hepburn Act amendment to the Interstate Commerce Act in response to public outrage over Standard Oil Company's allegedly abusive and discriminating use of pipelines to create a national monopoly in petroleum. The law designated all interstate oil pipelines as common carriers and granted the Interstate Commerce Commission (ICC) jurisdiction over their rates. Congress, however, excluded interstate carriers of water and natural gas from the common-carrier mandate and from federal oversight of transportation.

The issue of designating gas pipelines as common carriers was far from settled. In 1914 and especially following the surge of gas-pipeline construction in the late 1920s, that question prompted congressional hearings, either on its own merits or as part of a broader topic of inquiry.

In 1928, for example, the Senate passed a resolution requesting the Federal Trade Commission to survey the problems created by an epidemic of electric and gas-utility mergers. Corporate agglomeration had reached the point that more than three-fourths of the 50,000 miles of interstate natural gas pipelines were controlled by eleven holding companies—which also dominated gas production, distribution, and even electricity sales in many urban areas. (See Chapter 8 for a more detailed discussion of this era of the gas industry.) The commission issued its ninety-six-volume report in 1935, urging Congress to dismember these powerful corporations (Federal Trade Commission 1935). In addition, the report took a shot at the interstate gas transmission industry, suggesting that common-carrier requirements and federal regulation of pipeline services and rates were timely.

The New Deal Congress acted promptly, and the Public Utility Holding Company Act became law that same year. In the interest of doing something to tackle the immediate crises engendered by unrestrained growth of holding companies, Congress decided to leave the question of regulating interstate gas sales for another day. That day came three years later, with unanimous passage of the Natural Gas Act of 1938 (the NGA).

The act instituted federal oversight of rates charged by interstate gas-transmission companies. The Federal Power Commission, which owed its existence to the Federal Water Power Act of 1920, became the administering agency. In addition to rate regulation, the FPC held limited franchising powers. Nobody could build an interstate pipeline to deliver gas into a market already served by another gas pipeline without first obtaining FPC approval. In 1942, an amendment rounded out those powers by requiring commission certification of facilities penetrating new markets as well.

Given the weight of history, it now seems surprising that the 1938 act was endorsed by practically everybody. Although the impetus for passage came from the gas-importing states and their public utility commissions (which had been thwarted in trying to regulate end-use prices), the contested issues were compromised sufficiently to enlist support from interstate pipelines, gas producers, and gas-exporting states.

By 1938, interstate pipelines had seen the writing on the wall. Having succeeded in striking what they considered to be the most onerous provisions of the Federal Trade Commission recommendations (especially the common-carrier requirement and inclusion of direct industrial customers in the scope of rate regulation), the final version of the Natural Gas Act got their support. Interstate pipelines and remote producers eager to sell their gas recognized, too, that something was needed to bolster investor confidence in new pipeline construction now that internal financing through holding companies was on the wane. The franchising provisions of the bill offered some tangible assistance.

But the question of form and content for this entirely new area of federal regulation was so controversial that some ambiguous language was carried through to the final version of the Natural Gas Act. Congress, for example, left it to the implementing agency to determine the details of what constituted "just and reasonable" rates for pipeline services. Under the dictates of the 1898 *Smyth v. Ames* decision (169 U.S. 466), which called for "a fair return on fair value," and especially given the leniency that had long characterized oil-pipeline regulation by the Interstate Commerce Commission, gas-pipeline companies had little to fear. (The Federal Power Commission did, however, depart from the ICC ratemaking standard in the late 1940s by adopting original cost as the basis for pipeline valuation.)

The ambiguity that had the greatest historical impact was the NGA's treatment of sales from producers to interstate pipelines. Although the majority of Congress in 1938 was convinced of the need to establish a federal presence in the sale of gas by an interstate transmission company to a local distributor, there was an impasse on whether to extend federal authority upstream to reach wellhead transactions. Conveniently, the language was unclear: "The provisions of this chapter shall apply to the sale in interstate commerce of natural gas for resale . . . but shall not apply to the production or gathering of natural gas." Those who wished regulation to penetrate the production sector could believe that the act did so. Those who wanted to keep government out of the field could also view the act with favor.

## REGULATION OF WELLHEAD PRICES
## CHARGED BY AFFILIATED PRODUCERS

Soon after passage of the Natural Gas Act, the Federal Power Commission adopted a middle-ground position on the question of producer regulation. Affirmed by the Supreme Court in *Colorado Interstate*

*Gas Co. v. FPC* (324 U.S. 581 (1945)) and in *Interstate Natural Gas Company v. FPC* (331 U.S. 682 (1947)), the FPC ruled in 1942 and again in 1943 that field sales between a producer and an interstate transmission company would be subject to federal oversight if the two companies were *affiliated*. The commission reasoned that absent *arm's-length transactions* the pipeline company would have little incentive to bargain for a good price. Its utility status and the seemingly insatiable demand for this clean-burning fuel meant that a pipeline company would probably not suffer reduced demand if it agreed to excessive charges. What is more, the pipeline company had a powerful motivation to give the producer a good deal if intercorporate ties created an opportunity to evade the effects of pipeline regulation.

On the other hand, the commission felt that with respect to unaffiliated transactions, competition among *independent producers* in the field seeking buyers for their gas would be enough to hold wellhead prices in check. And if it was not, an affected party could seek relief under the longstanding provisions of the Sherman Antitrust Act that barred collusion and other anticompetitive activities. Indeed, the pipelines held substantial *monopsony* power over independent producers, which given the *cost-of-service* regulation of their own charges, tended to transfer income from producers to final consumers.

Especially in frontier areas like the Panhandle and Monroe fields, for many years there was far more gas available for sale than interstate pipelines could physically handle. Early field prices seemed to bear this out. Even absent wellhead regulation, pipeline companies were able to purchase gas at mere pennies per million btu, compared to coal-gas prices of several dollars.

Throughout the 1940s and 1950s, gas producers and even the FPC itself urged Congress to ensure that the commission's early rulings on affiliated and independent field sales would stick. The movement for congressional action strengthened following the Supreme Court decision on the *Interstate* case (referenced above) because the Court suggested that the arguments favoring regulation of affiliated field sales seemed to extend to independent sales as well, though that particular issue was outside the scope of the *Interstate* proceedings.

Congress twice passed legislation to clarify the act, explicitly exempting nonaffiliated field transactions from federal regulation. Both times, however, the bills met with presidential vetoes. In 1950, President Truman declared his objection to the amendment in no uncertain terms; he favored rate regulation of all field sales. Congress

tried again in 1956, this time prompted by a 1954 Supreme Court ruling that, absent congressional action, would compel the Federal Power
Commission to scrutinize independent field sales. The 1956 Congress
was joined, this time, by a president with a compatible ideology. But
President Eisenhower was nevertheless forced to veto the bill because
an attempted bribery of a Republican congressman by a producer
lobbyist lent an unsavory air to the position of the proponents. Although deregulation bills were periodically introduced, no legislation
emerged from Congress for another twenty-two years until passage
of the Natural Gas Policy Act.

## REGULATION OF ALL WELLHEAD PRICES

Since Congress declined to clarify its intent, the job of interpretation was left to the Federal Power Commission and ultimately to the
courts. The question reached the U.S. Supreme Court in 1954, at the
insistence of a consumer state. In its landmark decision, *Phillips
Petroleum Co. v. Wisconsin* (347 U.S. 672 (1954)), the court ruled
that despite sixteen years of contrary interpretation by the FPC, the
Natural Gas Act did indeed require federal oversight of all field sales
of gas destined for interstate commerce. The majority opinion was
held by six of the nine justices who participated in the *Phillips* case:

> There can be no dispute that the overriding Congressional purpose was
> to plug the "gap" in regulation of natural-gas companies resulting from
> judicial decisions prohibiting, on federal constitutional grounds, state
> regulation of many of the interstate commerce aspects of the natural-gas
> business. . . . Protection of consumers against exploitation at the hands of
> natural gas companies was the primary aim of the Natural Gas Act. At-
> tempts to weaken this protection by amendatory legislation exempting in-
> dependent natural-gas producers from federal regulation have repeatedly
> failed, and we refuse to achieve the same result by a strained interpretation
> of the existing statutory language.

Interestingly, two of the dissenting judges voiced these same
arguments, but in reverse. They maintained that states *are* able to regulate wellhead prices, and that if anybody is exploiting anybody else, independent producers are certainly on the downside:

> The states have been for over 35 years and are now enforcing regulatory
> laws covering production and gathering, including *pricing*, proration of

gas, ratable taking, unitization of fields, . . . if anything, the independents at the producing end of the pipelines were likewise the victims of monopolistic practices by the pipelines.

In interpreting the Natural Gas Act, the Court had a difficult time construing legislative intent. As one dissenting justice lamented, "The legislative history is not helpful." The Court, therefore had to infer intent based upon economic theory and market conditions concurrent with the 1938 Congress. But the events and ideological climate of the 1950s were powerful forces as well.

In 1954, the *burner-tip* price of gas in most sections of the country was roughly the same as its primary competitor: oil. But this did not mean that the producers were, necessarily, extracting unfair profits by charging "whatever the market would bear." By 1954, most of the gas being sold under new contracts came from nonassociated fields where the economic feasibility of gas production had to stand on its own. The average wellhead price for new gas contracts was about 11 cents per mmbtu. This compared with an average burner-tip price of around 50 cents. (See Figure 5–1.)

Close to four-fifths of the price paid by consumers, therefore, compensated regulated utilities for their transmission and distribu-

**Figure 5-1.**    Component Costs of End-Use Natural Gas Prices.

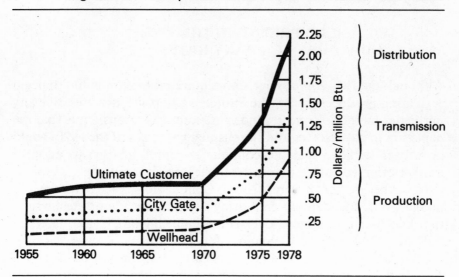

Source: American Gas Association.

tion services. In some parts of the country, the disparity was even more striking. Gas consumers in New York City, for example, paid $2.42 per mcf. Of that, only about 8 cents was attributable to well-head charges, with 24 cents going to the interstate pipeline while the distributor pocketed $2.10.

In retrospect, it seems that the consumer price of gas was close to that of oil prior to wellhead regulation not because producers were extracting unearned rents but because regulated utilities ate up whatever oil/gas differential might have existed, simply by extending service into outlying areas that were costlier to serve. Even after the Federal Power Commission began imposing wellhead-price ceilings, the same trend was apparent. And after the 1974 fly-up in world oil prices, transmission companies proposed all sorts of costly schemes for generating supplemental-gas supplies, including load upgrading and expansion of underground storage, unregulated gas and LNG imports, high-btu coal gasification, and Arctic pipelines.

With or without wellhead-price regulation, history shows us that the drive to reach a *market-clearing price*—thereby keeping gas demand in check with available supplies—is truly inexorable. Even while producer prices were held in check, a host of costly additions to utility systems were ready to fill the gap between the regulated rates and the prices consumers were willing to pay.

## WELLHEAD PRICING IN THE 1950s: INDIVIDUAL COMPANY RATES

Although the 1954 *Phillips* decision required the commission to begin regulating producer prices, the Supreme Court did not prescribe any particular methodology. In hindsight, some historians argue that the depletion in gas reserves and the ensuing gas crises of the 1970s could have been avoided had the commission simply chosen to adopt a market-oriented rather than a cost-oriented approach.

Nevertheless, the Federal Power Commission reacted to the Supreme Court mandate by employing the same regulatory techniques that had long governed its oversight of pipeline tariffs and, later, of affiliated field sales. The guiding principle was *cost-of-service* rate-making whereby companies are allowed to charge prices sufficient to recover the actual costs of providing gas, including a fair profit. Cost-of-service regulation of interstate gas pipelines was handled on a case-by-case basis because the number of companies was small and costs varied enormously among them. Even today, the twenty largest pipeline companies account for 85 percent of the interstate gas sold nationwide.

Transferring this kind of regulation to the gas-production sector soon proved unworkable. Foremost was simply the problem of numbers. For every interstate pipeline there existed hundreds or thousands of gas wells under a variety of ownerships, presenting the commission with a workload that was several orders of magnitude greater than its historic tasks. (Today, for example, 12,000 gas producers operate 200,000 wells.) In 1959, out of 1,265 separate applications for rate increases filed by independent producers, the commission managed to act on only 240. By 1960, the backlog had grown to over 3,300. The situation was so appalling that a member of the transition team for president-elect Kennedy concluded that "the Federal Power Commission represents the outstanding example of the breakdown of the administrative process." (Landis 1960:54; 390 U.S. at 756–58.)

## WELLHEAD-PRICE REGULATION IN THE 1960s: AREAWIDE PRICING

Even before President Kennedy was sworn in and new, consumer-oriented commissioners appointed, the Republican-appointed FPC had scrapped the case-by-case approach and put in its place *areawide pricing* (24 FPC 547 (1960)). That policy initially divided the nation's producing provinces into five regions and established that the ceiling price for interstate sales of gas in any one region was to be consistent throughout. It also swept aside the different regulatory treatment of affiliated and nonaffiliated field sales (although that distinction was resurrected following enactment of the Natural Gas Policy Act in 1978). Instead of several thousand determinations, the commission was now confronted by only five, and supporters of regulation were optimistic that this generic approach would prove to be manageable, efficient, and perhaps even fair.

The next ten years were, however, sorely disappointing. The commission set interim rates in 1960 based on the average contract price for gas sold during the previous year in each of the five areas. But it took five years of hearings and deliberations (and 140 pages of justifying text) to come up with a number to replace the interim rate for gas in the first region considered: the Permian Basin (34 FPC 159 (1965)). The commission devoted another three years to its Southern Louisiana decision, and by the end of the decade, it had not yet ruled on ceiling prices for the remaining areas.

Throughout the 1960s, wellhead prices were essentially frozen at 1959 levels. Not only did interim rates determine gas prices for regions in which the FPC had not yet taken action, but they became self-fulfilling even in the Permian and Southern Louisiana areas. When the commission set rates for these two areas in 1965 and 1968 respectively, the figures were remarkably similar to the interim rates that had been in effect since 1960. The ceiling for "new" Permian gas was 16.5 cents and for old or "flowing" gas, 14.5 cents. This compared to an interim rate of 16 cents. Likewise, the ceiling for new South Louisiana gas was exactly the same as the 21-cent interim rate.

It is hardly remarkable that the new area rates seemed to be clones of the interim rates. 1963 was the "base year" upon which detailed investigations of the costs of Southern Louisiana gas production were conducted. By 1963, however, the 1960 price freeze was undoubtedly skewing drilling and production decisions. Why should anyone drill for or produce gas that cost more than the price set by the commission in 1960?

The whole concept of regulating wellhead prices on a cost-of-service basis was doomed from the start; case-by-case regulation was logistically impossible, and areawide pricing attempted to find a norm where indeed none could exist. Actual costs for gas production varied enormously within a single region. One gas producer might have made a discovery with the first wildcat well; another producer might have punched a dozen dry holes before finding anything. One company might have hit gas at a thousand feet, and another may have drilled ten times deeper. Those who tapped gas reservoirs tainted by impurities like carbon dioxide and sulphur compounds had to sink extra capital into processing facilities before their gas was marketable. Moreover, there was and still is no universally accepted solution to the enigma of allocating a company's *joint costs* between oil and gas operations. As long as cost-based pricing prevailed, however, regulators were forced to devise some defensible formula for determining the costs of production when gas is found in association with oil. Finally, the commission was faced with incorporating into the ceiling price an amount that would adequately compensate gas producers for their endeavors. But what was a fair profit, and how could the commission calculate the *cost of capital* for a line of business that suffered risks far greater than those faced by public utilities?

Overall, then, the task that confronted federal regulators as a result of the Supreme Court *Phillips* decision was of a "damned if you do, damned if you don't" variety. Physically, the FPC could not do justice to field pricing on a case-by-case basis. The thousands of cases backlogged in 1960 left no doubt about that. Yet by turning to an areawide approach, there was no objective and even remotely "fair" approach to accomplishing the judicial mandate.

## SUPPLY-SIDE ACTIONS BY THE FPC
## DURING THE 1970s GAS SHORTAGES

### Adoption of a National Ceiling Price

Economists of various schools always have and probably always will argue about the best way to distribute *economic rent*. If a commodity can garner a price that far exceeds its actual costs of production, including a reasonable return on company investment, should the producer be allowed to reap an "unearned" windfall? Alternatively, would a greater social good be accomplished if consumers were favored via price controls, or should perhaps the government contrive a tax that would transfer the rent to taxpayers at large?

By the late 1960s, industry officials and some respected economists were arguing that the wellhead-price freeze instituted at the beginning of the decade was doing more than simply redistributing economic rent; prices were so low that regulation was, in fact, restraining the nation's capacity to find and produce gas. In 1967, and for the first time since the interstate gas industry was born, the nation's reserves began to decline—and continued to slide until 1981, when the producer price incentives of the Natural Gas Policy Act of 1978 began to work their course. Moreover, with gas demand on the rise, the *reserves-to-production ratio* was dropping precipitously below the historic standard of around 20. (See Chapter 6 for more detail on our nation's history of gas production.)

Whatever the actual effect of wellhead price regulation during the 1960s, there is no question that continued price restraint during the 1970s had some profound effects. This became especially apparent when burner-tip prices for gas failed to keep pace with OPEC-induced increases in oil prices. With every boost in domestic oil prices, consumers found gas just that much more attractive, and demand consequently surged. But because of price controls, producers could neither charge more for their existing supplies (thereby limiting customer demand to what was available), nor had they sufficient incentive to go out and look for more. What is more, with the prospect of deregulation just around the corner, consuming-state politicians and even the FPC suspected that producers were cutting back on the amount of gas actually delivered from reserves dedicated to interstate commerce. It was only a matter of time, therefore, before a shortage developed.

The problems posed by dwindling reserves and increasing demand were exacerbated by a growing schism between inter- and intrastate gas purchases. Intrastate customers could and would offer higher prices than interstate pipelines were allowed to pay. In 1965, only a third of the nation's total gas reserves were dedicated to intrastate markets; ten years later, almost half of the nation's proved reserves were committed to customers in the producing states. A more alarming statistic was the fact that prior to the NGPA, intrastate customers were able to outbid interstate pipelines for virtually all new gas discoveries shoreward of the federally owned Outer Continental Shelf, which, by definition, entered interstate commerce when pipelined to shore. (See Figure 5-2 and 5-3.)

The combination of supply and deliverability disincentives, the widening gap between end-use prices for gas and alternative fuels, and the purchasing-power advantage of customers within the major producing states—all induced by continuing wellhead price regulation—took its toll in the 1970s. The record-breaking cold winters during the middle of the decade made the situation even more unpleasant.

Again, economists of various schools (and politicians) advocate markedly different strategies for dealing with an impending shortage. On the one hand, if gas prices are deregulated and thereby can hold supply and demand in balance, a shortage is avoided altogether—but consumers have to pay more. On the other hand, if prices are held below a *market-clearing* level, some consumers do pay less, but others are shut out of the market completely or live in fear of future curtailments. Moreover, if price is not allowed to balance demand, then some governmental entity has to step in and decide who deserves to get how much of the limited supply.

**Figure 5-2.**   Inter- and Intrastate Gas Reserves.

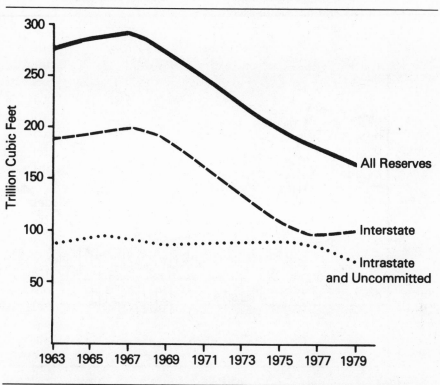

Source: (Adapted from) Energy Information Administration, U.S. Department of Energy, *Intrastate and Interstate Supply Markets under the Natural Gas Policy Act* (Washington, D.C.: U.S. Government Printing Office, October 1981).

This dispute was the prime reason it took so long for Congress to act even after the disastrous gas shortages that racked the Midwest and Northeast in the mid-1970s. With consumers already burdened by rising oil prices, it was at first unthinkable for our nation's leaders to allow our second-most important fuel—and a fuel that was almost entirely domestic in origin—to have its prices in effect dictated by the workings of a foreign cartel.

Throughout the 1970s, the majority position of FPC commissioners (whose five-year tenure depended upon presidential appointment and Senate confirmation) was indeed consistent with this general mood. Not until the winter of 1970–71, when 100 bcf of firm gas commitments were curtailed, did a new, Republican-dominated commission

**Figure 5–3.** Comparison of Annual Reserve Additions for Inter- and Intrastate Markets.

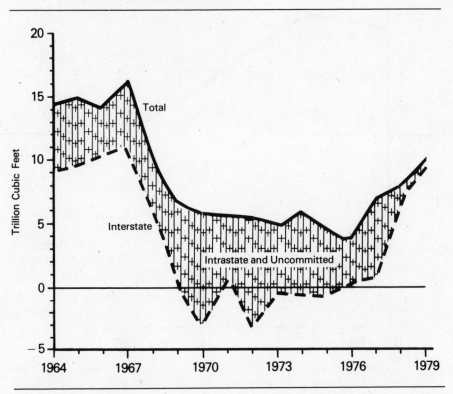

Source: Energy Information Administration, U.S. Department of Energy, *Intrastate and Interstate Supply Markets under the Natural Gas Policy Act* (Washington, D.C.: U.S. Government Printing Office, October 1981).

boost the prices that had prevailed throughout the 1960s. In June 1974, following the Arab oil embargo and the consequent jolt in oil prices, the commission abandoned areawide pricing altogether and moved to a single national standard (FPC Opinion No. 699). The new price formula no longer strictly adhered to cost-based utility principles. Market value, intrastate supply competition, and the prices of alternative fuels were additional considerations. Adoption of a uniform price also offered regulators a chance to respond expeditiously to changing economic and political realities.

The new *national price ceiling* of 42 cents per mcf doubled the rates that had prevailed during the 1960s, but it was still only half the price Texas and Louisiana producers were then receiving for gas sold intrastate. (Figure 5–4). Although the commission periodically increased price ceilings thereafter (Opinion No. 700 was issued in July 1976 and was subsequently revised in Opinion No. 700–A), it was politically barred from raising them enough to put much of a dent in the supply/demand imbalance. After all, Congress had continually considered bills to deregulate field prices, and it repeatedly rejected the idea until a compromise position was reached in 1978 with the Natural Gas Policy Act.

**Figure 5–4.** Comparison of Inter- and Intrastate Wellhead Prices.

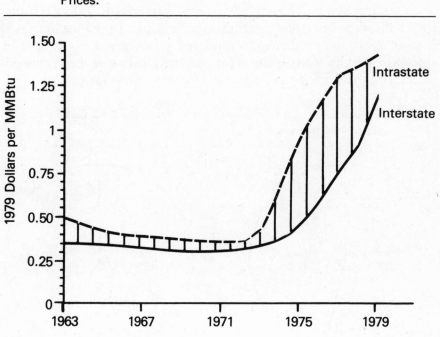

Source: Energy Information Administration, U.S. Department of Energy, *Intrastate and Interstate Supply Markets under the Natural Gas Policy Act* (Washington, D.C.: U.S. Government Printing Office, October 1981).

### Advance Payments, Emergency Purchases, and Self-Help Programs

In addition to boosting wellhead price ceilings, the commission attacked the market imbalance in less direct ways. It found several avenues for enticing gas out of intrastate markets, which, free of price controls, had been immune to the gas shortages that wracked the Northeast and the Great Lakes regions. While even schools in some northern cities were forced to close their doors during the height of the winter crises, electric-utility boilers in Texas were still gulping down gas. Consumers with access to intrastate supplies were literally awash in gas. In 1978, intrastate markets (primarily just the three states of Texas, Louisiana, and Oklahoma) accounted for 41 percent of annual gas sales nationwide and 47 percent of the remaining, dedicated gas reserves. (Figure 5–5.)

To combat the purchasing advantage of intrastate customers, the FPC authorized interstate transmission companies to channel interest-free loans to independent and affiliated producers if the producer agreed to sell the fruits of its efforts to the pipeline at the regulated price. The loan was to be reimbursed over a five-year period coin-

**Figure 5–5.**    Comparison of Inter- and Intrastate Gas Production and Reserves in 1978.

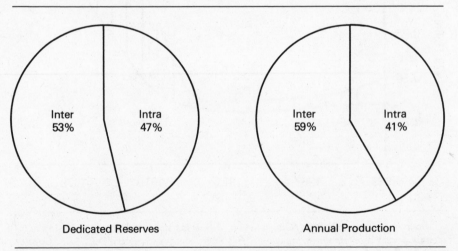

Inter 53%    Intra 47%        Inter 59%    Intra 41%

Dedicated Reserves                Annual Production

Source: Energy Information Administration, U.S. Department of Energy, *Gas Supplies of Interstate Natural Gas Pipeline Companies, 1978.* Washington, D.C.: U.S. Government Printing Office, April 1980.

ciding with the onset of gas production. If drilling proved unsuccess-ful, the loan was to be reimbursed in a similar manner, and under no circumstances could the producer wait more than five years before initiating the payback.

Meanwhile, the transmission company was allowed to treat these *advance payments* as part of the official ratebase from which cus-tomer charges were calculated. This meant that while a company might have sold bonds to raise the money, gas consumers ultimately were making the loan—and an interest-free loan at that.

The prospect of interest-free venture capital was, of course, attrac-tive to producers, and because the transmission companies could de-pend either on consumers or producers to provide a continual flow of cash to meet *sinking fund* obligations, advance payments boomed. Five years into the program, however, a U.S. circuit court (instigated by the New York Public Service Commission), directed the FPC to review the merits of advance payments to see whether gas consumers were receiving an adequate boom for their bucks. Following its review, the commission terminated the program as of December 1975, with outstanding commitments expected to phase out by 1981. When that decision was made, $3.3 billion had already been paid or irre-vocably committed, with another $2.2 billion tentatively dedicated.

With the advance-payments program, the FPC was only marginally successful in diverting new gas discoveries away from customers in producing states, but its track record with respect to old gas supplies was pretty good. Back in 1960 the commission had ruled that its au-thority to approve transmission company plans for abandoning fa-cilities extended to the supply contracts between an interstate pipeline and a producer. Regardless of the term that had been agreed on, a producer was not absolved from responsibility to continue supplying gas to an interstate pipeline even after the contract was defunct.

This ruling survived repeated attacks in the courts. Ironically, however, it might have brought more harm than good. It was, after all, just one more reason for producers to avoid committing gas reserves to an interstate company.

Just before the Natural Gas Policy Act of 1978 relieved federal regulators of much of the pressure to jerry-rig measures to combat the gas shortages, the commission took the initial steps to deal with the gas-deliverability problem. It warned that it would begin to review gas production activities from reserves dedicated to interstate commerce to ensure that deliveries were not unreasonably withheld.

The Department of Interior took more direct action; it imposed stipulations in Outer Continental Shelf leases and development plans that obligated companies to maximize delivery potential.

Perhaps the most successful supply-side efforts authorized by the FPC during the 1970s were the *emergency purchases* and *self-help* programs. Beginning in 1970, intrastate pipelines and producers were authorized to sell gas to interstate pipelines for periods of up to 60 days absent the usual federal oversight of the terms of sale. In 1973, the commission extended this sales window to 180 days, but following hostile reaction from authorities in some consuming states, it retracted the span to 60 days. The courts put an end to the program in 1975, ruling that the FPC could not shirk its mandate to regulate interstate "sales for resale" (*Consumer Federation of America v. FPC*, 515 F.2d 347 (D.C. Cir. 1975)).

Meanwhile in 1973, the commission instituted a self-help program whereby a retail distributor could go directly into the field and buy gas at prices higher than the federal ceiling price governing "sales for resale." The buyer would then arrange for an interstate pipeline to carry the gas on a contract basis instead of the usual system by which the transmission company sold both gas and its transportation services. The FPC ruled in 1975 that mainline industrial customers could also utilize contract pipeline services for self-help gas. The program survived litigation brought by transmission companies who felt threatened that it might ultimately nudge them out of the field altogether.

## CURTAILMENT POLICIES AND FPC DEMAND-SIDE ACTION DURING THE SHORTAGES

Despite the abandonment rulings, the increase in Outer Continental Shelf production, and the self-help, emergency purchases, and advance payments programs—all of which favored *inter*state sales—the shortage continued to deepen during the 1970s. Even moratoria on new customer hook-ups instituted by local distributors and their state regulators were not enough to keep supply problems from worsening.

First to feel the pinch were industrial customers and electric utilities in gas-importing states who had signed contracts giving their pipeline suppliers or distributors unilateral rights to cut off supplies if *firm* demand could not otherwise be met. These *interruptible* contracts

had been very attractive in an earlier era because they offered substantial price savings. Not all interruptible gas consumers were adequately prepared for cut-offs in supply, however, because pipelines
had for years maintained construction programs well ahead of anticipated demand. By the time the shortages hit, many interruptible
customers had been purchasing gas for a decade or more at bargain
rates without ever in fact experiencing an interruption.

When northern gas-utilities started shedding interruptible load
during the winter, there was no cost-free way for these customers to
make up the energy deficiency. Many had to invest in new facilities to
burn oil or coal that gave them a dual-fuel capacity. Those caught off
guard had to install equipment on an emergency basis. Even those
with existing dual-fuel capability found their costs escalating as they
were forced to operate oil- or coal-burning furnaces and boilers for
longer periods. By the time the shortages hit, the price of oil was far
greater than an equivalent amount of gas, and pollution-abatement
requirements for the cheapest, high-sulphur grades of residual oil
and coal posed additional costs.

As the shortage deepened, utilities found it necessary to cut back
on even their firm customers, and by doing so violated the terms of
those contracts. During the winter of 1976–77, interstate pipelines
were unable to fulfill about one-fourth of firm demand, with some
individual pipelines in considerably worse shape. Faced with litigious
customers, distributors appealed to their state commissions and interstate pipelines appealed to the FPC for help. Both bodies directed
gas-short utilities to *curtail* customers based upon a set schedule of
priorities—presumably reflecting state and national judgments of
need. Although the concept of government-directed curtailments
held up in court, some aggrieved customers have been awarded dam-
ages for instances in which a pipeline erred in rigorous implementation of an approved "curtailment schedule" or in which the customer
was severely injured. In the early 1980s (amidst a growing gas glut),
some of these cases were still making the judicial rounds.

Federal involvement in pipeline *curtailment schedules* began in
1971 when the FPC issued general guidelines. In 1972, the commission approved the first curtailment plan, which was filed by El Paso
Natural Gas Company. That plan created a five-tiered hierarchy of
service commitments. The following year, the commission prescribed
a generic curtailment plan consisting of nine categories for "priorities
of use." Although the generic plan was upheld in court, the judicial

branch ruled that the FPC, nevertheless, had to conduct separate inquiries and allow opportunities for public comment on curtailment schedules developed by each interstate pipeline company.

## CONGRESS TAKES THE LEAD: THE NATURAL GAS POLICY ACT OF 1978

Still unable to reach an accord on a long-term solution to the gas shortages, Congress took action in early February 1977 to deal specifically with that winter's energy crisis. With more than 4,000 manufacturing plants idle for lack of gas, a million workers laid off, and hundreds of schools closed, the Emergency Natural Gas Act (Public Law No. 95–2) gave the president extraordinary powers to intervene in private-sector decisions affecting gas supplies. The president (or his delegate, the FPC chairman) could order an interstate pipeline or a distribution company to relinquish some of its gas to a system whose customers were in worse straits. He could even grant special allowances whereby intrastate pipelines could carry interstate gas without becoming subject to federal tariff rulings. Then too, presidential proclamation could sanction "emergency" purchases of intrastate gas by interstate companies at prices above the FPC ceiling and free of FPC oversight.

Among the activities made possible by the Emergency Natural Gas Act was domestic use of "non-Jones Act" tankers normally restricted to international trade. Tankers dedicated to the South Alaska-Japan circuit made special trips, delivering liquefied natural gas from Alaska to Columbia Gas Transmission's system on the East Coast. El Paso Natural Gas Company, serving southern California, was able to transfer some of its western gas to needier customers in the East by making transportation arrangements with a string of intermediate carriers, including Delhi Gas Pipeline Company, Lo Vaca Gathering Company, United Gas Pipe Line Company, and Transcontinental Gas Pipe Line Corp. Congress was willing to extend the bounds of presidential discretion because the act was short term. Key provisions of the Emergency Natural Gas Act expired only a few months after the date of enactment.

Finally, in November 1978, Congress passed the Natural Gas Policy Act (NGPA) (Public Law No. 95–621), and its companion legislation, the Power Plant and Industrial Fuels Use Act (FUA) (Public

Law No. 95–620). By then, the proponents of a free-market price for natural gas had gained the edge, both in the White House and in the Senate. There was no end in sight to the gas shortage, and national leaders feared that if struck with another winter as cold as that of 1976–77, the political heat could become unbearable.

An increasing number of consumer-oriented congressmen faced up to the fact that wellhead regulation initially put in place to protect consumers had now gone full circle. Not only were potential users denied service altogether, but even existing customers with rights for firm service had to cope with shortages or, at best, live in fear of a cut-off. What is more, the nation was beginning to realize that relaxing price controls, thereby increasing domestic supplies, could back out OPEC oil, which was expensive, insecure, and a drain on the U.S. balance of payments.

The gas distribution companies (which traditionally have been passionate defenders of wellhead-price controls) divided on the issue, and a number of gas-utility executives joined the deregulation lobby. But the proponents of higher wellhead prices were able to secure the votes of reluctant congressmen only by compromising on the extent and pace of decontrol and by accepting broader federal authority over gas allocation. Those accommodations alienated many of the key constituencies in producer states. The final versions of the NGPA and the FUA were such hodgepodges of proconsumer and pro-producer rules that they were supported wholeheartedly by almost no one. Moreover, the legislation attracted an unlikely coalition of opponents from both importing and exporting states, which was only narrowly overcome by tough lobbying on the part of the Carter administration.

The detailed and prescriptive provisions of the 1978 legislation represented a reversal of philosophy from that of the 1938 Natural Gas Act, which had granted the Federal Power Commission wide discretionary authority. The NGPA created twenty-six categories of gas according to their *vintage* (the year in which a given gas supply was first committed to sale) and on the basis of a number of other production-related factors or characteristics of the producing company. Each category had its own statutory ceiling price and schedule for price escalations.

Perhaps it was fitting, therefore, that the Federal Power Commission by then was defunct. Instead, the Federal Energy Regulatory Commission (FERC) became the administering agency. FERC was born in 1977 when Congress passed the Department of Energy Organization Act (Public Law No. 95–91). With the exception of export/import authorizations (which were assigned to the Economic Regulatory Administration (ERA) within the new Department of Energy), FERC assumed all of the FPC's jurisdiction. FERC also took over oil-pipeline regulation, formerly a function of the Interstate Commerce Commission.

The NGPA provided for partial and phased deregulation of natural gas sold into interstate commerce. Deregulation was "phased" because, with the exception of *deep gas* (below 15,000 feet), which immediately was exempt from price controls, wellhead ceilings were to be relaxed gradually until 1985, when all price controls would disappear. Deregulation was "partial," however, because *old gas* brought into production before the date of enactment would forever remain subject to the price ceilings and escalation schedules dictated in the NGPA.

While the NGPA ultimately rolls back the reach of federal jurisdiction of interstate field transactions, the allocation provisions inserted to induce consumer-state support extended federal control—including control over wellhead prices—into intrastate markets until 1985. This encroachment on state powers was justified (and ultimately upheld in the courts) on the ground that intrastate sales did indeed "affect" interstate commerce.

In addition, the NGPA and the FUA together extended the reach of federal pricing and allocation policies further into end-use markets. The *incremental pricing* provisions of the NGPA loaded the bulk of higher gas costs onto industrial consumers. The FUA contained a number of *off-gas* provisions to reduce the amount of oil and gas consumed by large industrial plants and electric utilities.

The FUA became an anachronism almost before the ink had dried, however. Congress had failed to recognize that the gas shortage was an artifact of wellhead price regulation and consequently misjudged the relative availability of oil and gas. The ERA, which administered FUA, soon had to find administrative means to nullify the provision that would have forced conversions from gas to oil. In 1981, Congress included in its Omnibus Budget Reconciliation Act a FUA amendment that withdrew existing power plants from the act's scope.

With no end in sight to the North American gas glut and with swelling domestic and Canadian concerns about the linkages between high-sulphur coal, residual oil, and acid rain, the FUA looked increasingly out of step with the times. The NGPA, likewise, began to appear at odds with economic reality.

## GAS-MARKET UPHEAVALS OF THE EARLY 1980s

One problem with the NGPA became apparent following the 1979–80 spurt in OPEC oil prices. The act's sponsors had intended to accomplish a smooth phase-out of wellhead-price controls by 1985; however, the NGPA locked in a price-parity target keyed to an anticipated world crude-oil price of about $15 per barrel (in 1978 dollars). In 1981, crude oil was selling for almost $40 (about 34 1978 dollars), and virtually all government and industry officials had come to believe that its price would continue to soar without limit. As a result, they began to worry about a sudden gas-price "spike" or "fly-up" in 1985.

By 1983, the concern about a mid-decade price fly-up had vanished. The fly-up had, in fact, already appeared and virtually run its course. Crude-oil prices on the spot market had fallen to around $30 by late 1982, and in March 1983 OPEC brought its official price down to spot-market levels in the hope of forestalling an uncontrollable price collapse. But at the very time oil prices had been falling, gas prices had shot up.

Gas wells were now capable of producing and pipelines were committed to take more gas than consumers were willing to purchase at the going rates. An estimated 2 tcf per year of domestic gas was shut-in for lack of markets. Importers of Canadian gas cut their takes to less than half of authorized volumes.

A number of factors contributed to the switch from shortage to surplus. The most conspicuous was the eagerness of interstate gas pipelines—fresh from the 1970s curtailment traumas—to sign up for high-cost and deregulated gas with little regard to price. The immediate pre-NGPA era had also spawned a handful of big supplemental gas projects like LNG import schemes, which, by the early 1980s, were also inflating customer bills. Canada and Mexico, too, were selling gas at prices higher than "marginal" industrial customers were willing to pay.

## WHAT NEXT?

In reviewing the last ten years of gas regulation and market conditions, history reveals a confusing swing from shortage to surplus. If we have learned anything from the past decade, it is that the unforeseen impacts of complex legislation are capable of overwhelming its intended effect. The attempts of government to structure markets according to some preconceived plan are likely to fail completely when, as was the case with the federal administration and Congress in 1977–78, the planners misunderstand the market forces they are attempting to channel.

Nevertheless, in considering the history of the gas industry and its regulation over the longer term, it becomes apparent that "we have been here before." All of today's burning questions (and, most likely, tomorrow's) have their counterparts in the past. Moreover, the creep of government authority from one *regulatory gap* to the next was logical and, in hindsight, utterly predictable. With passage of the Natural Gas Policy Act and the Fuels Use Act in 1978, Congress extended the federal reach to perhaps the inevitable extreme. With intrastate transactions and end-user fuel choices now subject to federal oversight, virtually no regulatory frontiers remain.

Ironically, the crack in regulation opened by the NGPA and the forces set in motion by the regulation-induced gas shortages (including load-upgrading, high-priced gas imports, and expensive supplemental-gas projects) conspired to drive final consumer prices to levels at least as high as those that would have prevailed without regulation. By 1983, it was hard to identify any long-term consumer benefits that would flow from continued regulation of wellhead prices, end-uses, and even interstate gas-transmission tariffs.

The regulatory approach to problem solving in the gas industry had thus reached the end of its string. Even those interest groups and public officials who are normally distrustful of big business (and who are comfortable only with economic decisions for which some public entity is accountable) are now hard-pressed to find any plausible new combination of regulatory measures that offers hope for rolling back consumer prices. The internal logic of gas-market behavior, which for decades pressed inexorably for ever-wider government control, now points toward a devolution of federal authority.

Whether and when further decontrol of the production, end-user, transmission, and distribution sectors is likely to take place is a question that demands a far broader understanding than this survey of regulatory history provides. The outlook for future gas discoveries is an essential consideration. So too is an awareness of how gas markets function today and how they would likely operate in a less regulated environment. The next several chapters attend to these questions.

## REFERENCES

Federal Trade Commission. *Summary Report of the Federal Trade Commission to the Senate of the United States Pursuant to S.R. 83 on Holding and Operating Companies of Electric and Gas Utilities, No. 73-A*. Washington, D.C.: U.S. Government Printing Office, 1935.

Landis, James M. *Report on Regulatory Agencies to the President-Elect*. P. 54 or 390 U.S. at 756-758.

# 6 A WANING RESOURCE?
## Gas Production in North America: Its History and Outlook

One of the unfortunate characteristics of natural gas (and, for that matter, any resource underground) is the high cost of confirming its existence at any particular location and at any particular depth. No company is likely to go out looking for hydrocarbons unless it has a good chance of finding a buyer at a price that justifies the development costs. While this notion is hardly revolutionary, it is sobering to review just how often during the last hundred years public and private doomsayers have interpreted a depletion of known supplies as irrefutable evidence of absolute resource exhaustion.

During the 1970s, informed observers realized that the decline in U.S. gas reserves was in large measure a result of archaic wellhead-price controls. Nevertheless, the decline in U.S. oil reserves, even in the absence of a constricting regulatory environment, led many to believe that we would soon deplete our two most important energy resources.

The perception of imminent resource exhaustion is not new. In one form or another it has been with us since the eighteenth century, when population and economic growth were first recognized as long-term trends rather than random fluctuations. Over the last century, geologists and engineers in the United States have predicted an impending energy crisis nearly every generation, based upon the exhaustion of the dominant fossil fuel of the day, whether it was oil, coal, or gas. All of these forecasts were obviously premature, and some look ridiculous in hindsight.

The depletion of shallow Appalachian gas reserves in the early 1920s, for example, prompted all sorts of dire predictions and government programs to allocate remaining supplies. The hysteria was not quelled even by discovery of the Panhandle field in northern Texas, which, at the time, was the largest gas field known anywhere in the world. At the then-current state of pipeline technologies, Panhandle gas was no more accessible to dependent eastern users than if it had been on the moon.

Similarly, during World War I, the U.S. Geological Survey declared that the age of petroleum was over and that oil production would peak in the mid-1920s. This prediction was prompted by the realization that almost all of the naturally occurring oil seeps had been drilled. The more sophisticated geologists did point out that a great deal of crude oil undoubtedly remained, but conventional technologies offered no hope for locating those reserves by any method other than hit-or-miss random drilling throughout the nation's sedimentary basins, at an unacceptably high cost.

Seismic exploration techniques were not unknown when the experts concluded that for all practical purposes our nation's endowment of oil was almost gone. But these methods did not become commonplace until the 1920s—that is, until economic necessity sparked the transformation of scientific knowledge into industrial innovation. By 1930, however, geophysicists had instigated a revolution in petroleum exploration, and the reserves discovered since the date on which U.S. oil supplies were first forecast to run out are many times the volumes found previously.

At some level of generalization it is true that depletion of nonrenewable energy resources is an absolute and irreversible phenomenon. But absolute depletion of the earth's hydrocarbons is far from imminent. Each of three individual deposits in the Western Hemisphere—one in Alaska, one in Canada, and one in Venezuela—contains heavy hydrocarbons in volumes comparable to the whole world's proved, probable, and speculative reserves of crude oil as conventionally defined. Moreover, gas contained in the *tight* Devonian shales of the Appalachian region or the *geopressurized brines* of the Gulf Coast states could, if produced, support current levels of domestic gas consumption for hundreds or even thousands of years. Methane *hydrates* (gas frozen into a physical-chemical bond with water) under the sea and in the Arctic constitute another huge and so far undeveloped gas resource.

None of these deposits is included in official estimates of proved and probable oil or gas resources. At the present time, neither the world's energy hunger nor technological sophistication has reached levels at which these unconventional resources can be viewed as saleable commodities. Twenty years from today, however, our skepticism about the practical worth of these reserves may look just about as foolish as the doomsday predictions of the 1920s now appear.

## FUNDAMENTALS OF PETROLEUM GEOLOGY

Projecting future gas supplies is truly a committee effort, wedding the best guesses of petroleum geologists, applied technologists, economists, and political pundits. It is far more an art than a science. What is more, history seems to tell us to take any such predictions with a grain of salt.

Not only are gas-supply projections almost always proved wrong, but they tend to become outdated even before they are published. For both these reasons, in this text we will neither venture our own forecasts nor include those of anyone else. We will, however, set forth relevant facts and theories in order to help the reader understand the long-term supply outlook as it appears today and as it may change in the future.

### The Importance of Sedimentary Basins

One thing on which practically all geologists agree is that the best places to explore for hydrocarbons are regions that at one time or another supported lowlying environments or shallow seas that trapped sediments from higher ground. Compressed into rock and sometimes folded and faulted into various contortions, these areas are nevertheless called *sedimentary basins*.

Oil and gas can be found only in rocks whose solid constituents are not tightly packed together. Of the three kinds of rocks that comprise the earth's crust, (sedimentary, igneous, and metamorphic), sedimentary rocks are most apt to contain pores because, by definition, they represent accumulations of eroded rock particles or precipitates brought together by the actions of wind, water, and gravity. The richest hydrocarbon-bearing strata are usually sandstones because particles of sand have a tendency to settle together like little ball bearings, thereby leaving room for fluids to accumulate.

Sandstones are thus the best *reservoir* rocks because they are usually permeable as well as porous. High permeability means that the pores are well-connected so that gaseous and liquid fluids can readily flow in response to gravity or changes in pressure. Good examples of strata that are rich in hydrocarbons but of poor permeability are the Devonian shales of the Appalachian region and the *tight sands* of the Rocky Mountain states. In both of these regions, the flow from a well is so slow that the Natural Gas Policy Act of 1978 provided an incentive pricing scheme. In 1982, for example, the act allowed producers of Devonian shales and tight sands to charge twice as much as what producers of conventional gas could receive.

One additional feature that characterizes all hydrocarbon reservoirs is the presence of a trapping mechanism. Unless an impermeable rock *cap* seals the gas-bearing strata, the fluids will escape, moving outward and upward under the tremendous weight of hundreds or thousands of feet of overburden. For this reason, petroleum geologists look for *anticlines* or other structural features that resemble inverted containers within the world's 600 sedimentary basins. The sedimentary basins in the United States are depicted in Map 6–1.

### Biotic versus Abiotic Theories of Hydrocarbon Genesis

The foregoing discussion of sedimentary reservoir rocks does not, however, explain how the hydrocarbons get there in the first place. No consensus unites the scientific community with respect to the genesis of hydrocarbons. In fact, as the years go by, the question seems even more puzzling. New evidence erodes even the foundations of traditional theory because gas is turning up in places where conventional wisdom maintains it should not exist.

The theory carrying the most respected pedigree holds that natural gas, crude oil, coal, and other hydrocarbons are almost entirely of biological origin. That is, the hydrogen and carbon atoms within those substances once comprised the tissues of living organisms.

**Map 6-1.**   Major Sedimentary Basins of the United States.

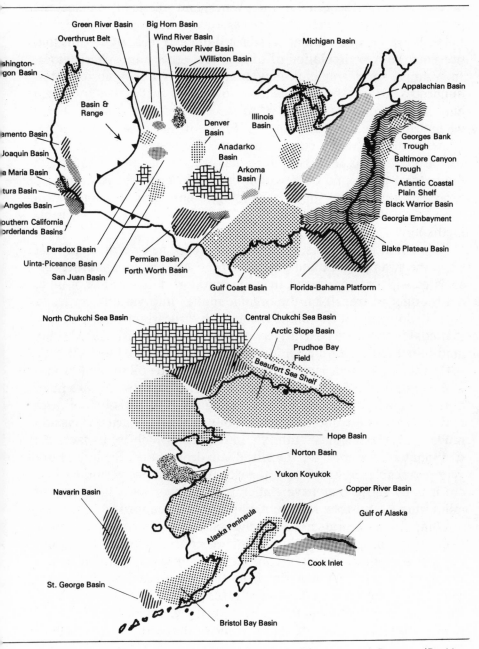

Green River Basin
Big Horn Basin
Overthrust Belt
Wind River Basin
Powder River Basin
Williston Basin
Michigan Basin
shington-
gon Basin
Appalachian Basin
Basin &
Range
Denver
Basin
Illinois
Basin
amento Basin
Georges Bank
Trough
Joaquin Basin
Anadarko
Basin
Baltimore Canyon
Trough
a Maria Basin
Arkoma
Basin
Atlantic Coastal
Plain Shelf
tura Basin
Angeles Basin
Black Warrior Basin
outhern California
orderlands Basins
Georgia Embayment
Paradox Basin
Blake Plateau Basin
Uinta-Piceance Basin
Permian Basin
Forth Worth Basin
San Juan Basin
Gulf Coast Basin
Florida-Bahama Platform

North Chukchi Sea Basin
Central Chukchi Sea Basin
Arctic Slope Basin
Prudhoe Bay
Field
Beaufort Sea Shelf
Hope Basin
Norton Basin
Yukon Koyukok
Navarin Basin
Copper River Basin
Gulf of Alaska
Alaska Peninsula
Cook Inlet
St. George Basin
Bristol Bay Basin

Source: Adapted from Riva, Joseph P., *World Petroleum Resources and Reserves* (Boulder, o.: Westview Press, 1983).

Most organisms upon death suffer the normal consequences of decay. Oxygen present in the wind, water, and soil slowly oxidizes the hydrogen atoms into water and the carbon into carbon dioxide. Sometimes, however, organic matter is buried under a mass of detritus that prevents the free circulation of oxygen and other atmospheric gases. In swamps, for example, organic matter that has been buried a few hundred years can become peat—a low-grade fuel whose hydrogen and carbon contents oxidize at an accelerated rate (that is, burn) when a flame sets off the reaction. Some of that organic matter becomes methane, which consists of a single carbon atom surrounded by four hydrogens. This *swamp gas* has exactly the same chemistry as the natural gas found deep within the earth, and it is generated by processes akin to the simple technologies now used in many urban areas to convert sludge, wood chips, and other *biomass* into useable methane.

Given millions of years, thousands of feet of new detritus may rest upon the swamp muck and peat. The pressures and temperatures associated with greater depths, in conjunction with the passage of time, are thought to transform the organic matter into various grades of coal. Proponents of this *biotic* theory of hydrocarbon genesis believe that coal is comprised exclusively of terrestrial detritus. Indeed, only land-based fossils have been identified within any coal deposit.

On the other hand, most experts attribute crude oil to shallow marine environments. A hundred million years from now, for example, organic matter in the detritus now piling up in the Mississippi delta may well become crude oil squeezed out of the compacted clays and muds. Clays and muds "lithify" into shale, which is, in fact, the dominant *source rock* for the world's hydrocarbons. Shales are not very permeable, however, and the richest oil and gas accumulations are usually those that have "migrated" tens or even hundreds of miles into a sandstone reservoir, appropriately capped.

What about natural gas? Methane—along with heavier *gas liquids* like ethane, propane, and butane—is of course associated with oil, but it is sometimes found in conjunction with coal. Methane is also found in isolated sediments far from the nearest coal or oil deposit. More surprising is the way that it turns up just about everywhere, from water wells to permafrost to volcanoes, giving rise to *abiotic* explanations for the origin of at least some of the earth's endowment of natural gas.

Methane, the simplest of all hydrocarbons, may be one of the original gases that comprised our infant planet. Indeed, the atmospheres of the outer planets contain vast amounts of methane. Perhaps much of the earth's original methane, and other primeval gases like carbon dioxide, nitrogen oxides, and hydrogen sulphides, were trapped as the crust congealed. And as the crust now heaves and groans and the earth's innards churn, those gases wend their way upward through cracks and fissures and from pore to pore.

That gas would tend to accumulate in the same places as the biotic theory would have it: porous sedimentary strata capped by impervious rock. No wonder the abiotic theory is so new and still widely suspect. With few exceptions, it is simply not needed.

But the evidence is mounting. Studies suggest that even earthquakes routinely unleash methane from the earth's bowels. Detection of methane and other gaseous emissions by animals with noses superior to our own has become a plausible explanation for the often-reported behavioral changes of pets and farmstock just prior to a quake. The abiotic theory is also the best explanation for traces of methane sometimes found trapped inside volcanic and other igneous rocks.

One of the most compelling pieces of evidence in support of the abiotic theory is the presence of gas found deeper within the earth than anyone had imagined possible. Technological and economic factors that currently limit our ability or simply dissuade us from drilling much beyond 20,000 or 30,000 feet do not even begin to test the bounds of potential deep gas resources. The cost of drilling an additional foot at 20,000 feet is ten times the cost of similar operations at the 1,000-foot level.

Another disincentive to probing the lower reaches of sedimentary basins (which may be 50,000 feet thick) is that a driller must be willing to look exclusively for gas. Oil becomes progressively scarcer at depths beyond 9,500 feet, and methane and gas liquids begin to dominate. Finally, at about the 16,000-foot gradient, any oil that might once have been present would have long cooked away, "cracking" into simpler methane molecules.

Finally, although gas found in *geopressurized brines* and *methane hydrates* is conducive to biotic explanations, the sheer magnitude of the resource and its distribution across the planet give further credence to abiotic theories. Gas truly does seem to pop up just about everywhere.

The biotic/abiotic debate is of more than academic importance. If the abiotic theory catches on, it could revolutionize the industry's approach to gas exploration and applied technologies. Perhaps, too, the relative availability of gas compared to crude oil, in conjunction with its clean-burning characteristics, might elevate gas to our number-one energy source.

In a fundamental way, gas already is more important than oil. Gas is often found and produced from reservoirs totally devoid of oil. Fully three-fourths of the known gas in the United States is *nonassociated* with oil. Oil, however, is rarely found without gas, and if it is, it might just as well not exist. For methane, as an expandable vapor, provides the motive lift that drives oil upward through a well in defiance of gravity. One of the great tragedies that accompanied the birth of the petroleum industry was the tremendous resource waste. Not only were huge volumes of natural gas vented into the atmosphere (probably upwards of 50 tcf have been flared in the United States), but much of the original oil in place was stranded as a result and is still inaccessible.

Natural gas also holds an edge over oil because, with the exception of tight formations of very low permeability, almost all of the resident methane will flow out of the well on its own volition. Gas recoveries of 80 or 90 percent are common without any type of enhanced recovery. On the other hand, only half or less of the in-place crude can usually be extracted, even using advanced production techniques like gas reinjection and waterflooding.

## DOMESTIC SUPPLIES OF NATURAL GAS:
## A HISTORICAL PERSPECTIVE

At year-end 1982, the official estimate of proved reserves of gas remaining in the United States was around 200 trillion cubic feet (tcf). At current consumption rates of about 20 tcf per year, that amount could furnish all U.S. gas needs for about ten years. To most industry observers, a *reserves-to-production ratio (R/P)* of ten, however, is considered to be dangerously low. Moreover, it represents the lowest level of gas inventories that the U.S. has tallied since reliable estimates first became available.

The official estimate of U.S. gas reserves hit an all time high of 290 tcf in 1967. For the next thirteen years, reserve additions trailed production rates. In 1981, however, the trend reversed. Posted additions surpassed production by 14 percent that year and by 11 percent in 1982. The upturn is universally attributed to the workings of the Natural Gas Policy Act, which in 1978 relaxed wellhead-price controls for conventional gas supplies, provided additional incentives for production of difficult reservoirs like tight sands, and freed deep gas to capture whatever price the market would bear.

One of the anomalies of the gas supply and demand situation in the early 1980s was that despite the low R/P ratio, the nation was in the midst of a gas glut. Interstate transmission companies found that they could not sell all of the gas that they were obligated to take. Many supply contracts, particularly those negotiated during the seller's market of the late 1960s and 1970s, included provisions that required a buyer to pay for 75 or even 90 percent of a property's gas *deliverability*— even if the pipeline chose not to take the gas because of slack demand. Several forces contributed to the *take-or-pay* crunch facing the gas transmission industry. The acquisition of deep gas at prices equivalent to crude oil at $50 or $60 per barrel called for enormous increases in customer billings. Overoptimism on the part of the industry about the value of gas, coupled with an unexpected downturn in fuel-oil prices, forced many industrial customers to flee to other fuels.

Meanwhile, high interest rates, coupled with continuing regulation of *vintaged* gas already discovered, meant that cash in the bank was a more profitable investment than gas in the ground. This meant that producers holding take-or-pay contracts with gas pipelines could make unilateral decisions to increase profits by boosting deliverability. Were it not for the fact that punching new holes to increase production is a costly activity in itself, the scramble to cash out reserves might have been much worse.

One other factor that contributed to the gas glut was that neither pipeline bondholders nor federal regulators were policing the new gas-purchase transactions. During the 1950s, supply contracts were usually issued in conjunction with plans for construction of a new interstate pipeline. Prospective bondholders demanded that enough gas be committed to the project to ensure that revenues would be forthcoming throughout the twenty- to thirty-year life of the bonds. The Federal Power Commission, as well, required dedication of adequate reserves before approving construction of new facilities.

This prerequisite for pipeline construction was so powerful that some producers found it necessary to enter into *warranty sales con-*

*tracts*, pledging to deliver a threshold volume of sales without limiting its scope to any particular field or fields. Warranty contracts caused trouble, however, when gas prices shot up in the early 1980s. In 1983, for example, a major gas producer in Louisiana, still bound by the terms of an old warranty contract, was purchasing gas from another producer at a price approaching $3.00 per mmbtu and then selling it to an intrastate pipeline at the warranty price of only 28 cents per mmbtu.

The financing imperatives for new pipeline construction that prompted dedication of reserves and warranty contracts in the past are no longer pressing. Since the 1970s, new supply contracts have generally replaced expired contracts. With few exceptions, the big construction era of the gas-transmission industry is over.

After wellhead-price deregulation has had a chance to reinvigorate gas exploration and production, and after the inevitable marketing quirks of the transition period are settled, how will the nation know when a healthy level of inventories has been restored? Because the gas industry was at its prime during the 1950s and because that decade displayed an R/P ratio exceeding 20 conventional wisdom may assume that twenty years' worth of inventories is ideal. But unless overall economic conditions closely resemble the 1950s (including 3 to 6 percent interest rates), an R/P goal of 20 may be so much nostalgic whimsy.

Moreover, the 1950s evidenced a period of gas oversupply, simply because interstate pipelines could not be built fast enough to connect willing sellers with interested buyers. An R/P ratio of 20, therefore, did not represent a stable and healthy level of inventories.

As the years rolled by, higher proportions of nonassociated gas came into production. (In 1950, over a third of U.S. production came from associated wells; by 1980 that statistic had fallen to 18 percent.) Nonassociated gas can be developed without any of the usual constraints placed upon associated gas for maintaining oil-reservoir pressures. The rate of production of a nonassociated gas field is limited only by the operator's interest in drilling a higher (and therefore costlier) density of wells. A 1982 report by the Congressional Research Service (Schanz and Riva 1982) observed that on a field-by-field basis, a reserves-to-production ratio of 10 or slightly lower was the norm for gas, compared to about 14 for oil. If the nation's combined R/P ratio for gas swells much beyond 10, therefore, it would mean that some companies are burdened with *shut-in* supplies or have held back on field investments so that deliverability is less than optimal. For all these reasons, nationwide gas inventories amounting to ten years' production at current rates may actually signal a healthy balance between supply and demand.

## THE OUTLOOK FOR DOMESTIC SUPPLIES OF CONVENTIONAL GAS

One of the big questions in the gas industry today is the extent to which deregulation of wellhead prices will bolster our nation's gas supplies, both over the short- and long-term. How much gas really is out there waiting to be discovered, and how much of what becomes known will be economical to produce under prevailing market conditions and state-of-the-art technologies? Specifically, what is the outlook for domestic supplies of conventional natural gas?

From the standpoint of sheer physical presence, the outlook for finding new gas reserves in the United States is decidedly better than it is for oil. In the early days of the oil business, gas was mostly just a bother, and huge volumes were flared. When technology advanced to the point where methane could be moved more than a hundred miles, the industry could not build pipelines fast enough to hook up old and newly discovered reserves. Just when supply and market outlook had a chance of reaching equilibrium, wellhead-price ceilings created wholly new disincentives. Were it not for the fact that gas resources are often found in association with oil, the frightening depletion in gas reserves and deliverability during the 1970s might have been much worse—or perhaps it might have prompted the nation to overhaul an antiquated regulatory system before the OPEC price upheavals turned a problem into a crisis.

The U.S. mainland may well be a mature province from the standpoint of opportunities for new discoveries of oil, but for gas it is still full of promise. What is more, the potential exists for exciting new discoveries of gas even in worked-over areas. (See Map 6–2 and Table 6–1.) The new regulatory climate makes it worthwhile to drill deeper than ever before imagined, probing strata (most notably in Oklahoma's Anadarko Basin and Louisiana's offshore Tuscaloosa Trend) far below the potential oil horizons. But even within the shallow zones, a good deal of gas may simply have been overlooked in the headlong quest for oil. For unless drillers specifically test for gas, rigs

**Map 6-2.**   Major Natural Gas-Producing Areas of the United States.

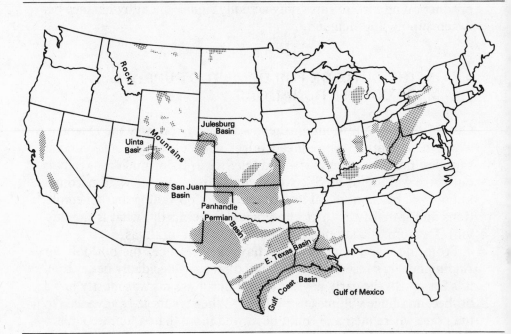

**Table 6-1.**   1981 U.S. Drilling: Leading Geologic Provinces (ranked by number of new field wildcats).

| Rank | Province | Oil | Gas | Dry | Total | Average Depth | Percentage Successful |
|------|----------|-----|-----|-----|-------|---------------|----------------------|
| 1 | Gulf Coast Basin | 58 | 160 | 981 | 1,199 | 8,988 | 18.2 |
| 2 | Permian Basin | 154 | 78 | 690 | 922 | 6,941 | 25.2 |
| 3 | Anadarko Basin | 81 | 72 | 481 | 634 | 6,233 | 24.1 |
| 4 | Central Kansas Uplift | 67 | 5 | 420 | 492 | 3,786 | 14.6 |
| 5 | Bend Arch | 44 | 29 | 413 | 486 | 3,619 | 15.0 |
| 6 | Williston Basin | 85 | 11 | 358 | 454 | 8,833 | 21.1 |
| 7 | Arkla Basin | 24 | 26 | 385 | 435 | 6,001 | 11.5 |
| 8 | Denver Basin | 63 | 20 | 347 | 430 | 5,606 | 19.3 |
| 9 | Mid-Gulf Coast Basin | 23 | 13 | 326 | 362 | 9,450 | 9.9 |
| 10 | Powder River Basin | 29 | 1 | 255 | 285 | 7,484 | 10.5 |

Source: Petroleum Information Corporation, *Resume 81: The Complete Yearly Review of U.S. Oil and Gas Activity,* 1982.

can push through strata rich in this odorless vapor without detecting any clues. Later, the well's casing effectively locks out the methane streams, along with undesirable fluids like groundwater.

## Gas Prospects in the "Overthrust" Belts

The industry is drilling for gas today in places where even seismic investigations would have been laughed at ten years ago. The *overthrust belts* of the Rockies and Appalachians are prime examples. Here, geologists are probing the sediments that underlie masses of rock heaved up and over the adjacent countryside by *tectonic* (mountain-building) forces.

The Western Overthrust Belt extends over 2,000 miles from Alaska into Central America. (See Map 6-3.) In the United States, exploration ventures targetted the northern region. In the early 1980s, rigs gathered in western Wyoming, spilling into Utah and Idaho and testing prospects from the Snake River south to the Uinta Mountains.

Until the late 1970s, the Overthrust Belt was not viewed as a rich hydrocarbon province and was, in fact, at one time considered nearly worthless. But between 1975 and 1982, nineteen important gas fields were discovered between the Canadian border and the Uintas, and the region became widely regarded as one of the nation's most attractive onshore provinces for *wildcat* drilling in untested strata. Much of that activity was prompted by the NGPA-induced incentives for deep drilling. Nevertheless, the area holds considerable promise for midrange discoveries that will likely become profitable when special incentives for deep gas are gone and the glut of the early 1980s has dissipated.

By mid-1982, the nation's most attractive deep-gas provinces were beginning to worry those producers that had made substantial investments. Independents and very small companies accounted for a large percentage of the exploratory action, and they were far less resilient than the majors in coping with market downturns. Encouraged by unrealistically high prices for deep gas (in the range of $7 to $10 per mmbtu), too many companies were drilling too deep. In 1983, nobody was buying deep gas for anywhere near the earlier prices, and some pipelines had altogether stopped shopping for new contracts. Even some of the producers with sales locked in by take-or-pay clauses found that purchasers had reduced their takes to the minimum legally allowed and were itching for price renegotiations.

**Map 6-3.**    Deep Gas, Devonian Shales, and Tight Gas in the United States

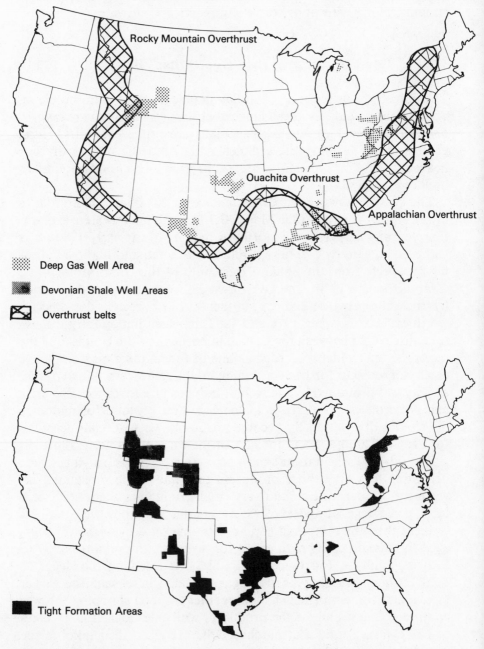

Source: Federal Energy Regulatory Commission.

Another problem that confronted gas owners in the Western Over-thrust Belt was the lack of means for getting their product to market. As soon as industry interest heightened, gas-transmission companies began planning for new pipeline construction. In the autumn of 1979, Cities Service Gas Company (now Northwest Central Pipeline Corporation) began moving gas through its new 20-inch diameter, 610-mile pipeline system, which stretched from Rawlings, Wyoming, to existing pipeline facilities in the Hugoton Field of Kansas. That pipeline was relatively easy to finance because it involved only 135 miles of new gas plant. Cities Service relied, instead, upon conversion of an existing crude-oil pipeline acquired from Arapahoe Pipeline Company.

Other pipeline proposals for connecting Overthrust gas, which relied on bigger economies of scale (larger pipe diameters) and totally new construction, were far more difficult to finance. *Joint ventures* backed by several existing transmission companies, therefore, sponsored most of the proposals for serving markets both east and west of the Overthrust Belt. Rocky Mountain Pipeline Company applied for a license to carry Wyoming Overthrust gas 580 miles south and west to southern California (at a projected cost of $575 million). Sponsors of the Pathfinder pipeline proposed a project of similar scale that would carry gas eastward to connections with existing midcontinent trunklines, and Trans-Anadarko sponsors envisioned an Overthrust connection as part of their North-Texas to Louisiana construction plans. But as of mid-1983, Trailblazer Pipeline System was the only consortium that had conquered all the regulatory hurdles, with construction of an eastward pipeline underway.

Federal approval of the 800-mile Trailblazer project was far from routine. The initial filing was made in November 1978 by Natural Gas Pipeline Company of America (now a subsidiary of MidCon Corp.), Columbia Gulf Transmission Company, and Colorado Interstate Corp. The sponsors silenced some of the initial opposition from smaller carriers in the producing areas (who advocated expansions of their own systems) by bringing Mountain Fuel Supply Company into the consortium. Nevertheless, opposition from Rocky Mountain Pipeline and Trans-Anadarko sponsors was intense. If reserves were inadequate to support all proposals, which of them would be of most benefit to the nation?

Skepticism regarding the adequacy of reserves delayed licensing for several years. Federal regulators wanted proof that enough gas was available to support the 36-inch Trailblazer system. Yet pipeline advocates argued that gas owners were delaying expenditures for field delineation until pipeline construction became more certain. Then too, some producers approached by Trailblazer as well as Rocky Mountain, Trans-Anadarko, and sponsors of other as-yet phantom facilities felt it best to delay commitments until the victor became apparent.

Trailblazer's financing plan and tariff also faced obstacles. Two other interstate carriers (Tenneco, Inc. and InterNorth, Inc.) joined the team, but sponsors still lacked the wherewithal to finance the half-billion dollar project on a conventional basis. Finally, Trailblazer was granted tariff terms that resembled some of the risk-transfer techniques developed during the days of LNG, coal gasification, and Arctic *supplemental* projects. Four years after filing their application, the Trailblazer sponsors broke ground.

By 1982, exploration activity in the Eastern Overthrust Belt had also intensified. Although the new geological wisdom supported a good deal of optimism, the Appalachian thrust zone has not yet produced anything substantial. The search began in earnest, however, only after wildcat successes in the western belt prompted the industry to look for similar geological structures elsewhere, and America's oldest producing area is still truly a frontier with respect to overthrust prospects. With 60,000 square miles of untested deep sediments, the Eastern Overthrust belt qualifies as one of the least-explored areas in the United States today.

A third thrust zone is also attracting industry interest. The Ouachita Overthrust Belt (centered in southern Oklahoma but extending eastward into Arkansas and southwesterly into Texas) has been rewarding wildcatters since 1977. As of 1982, however, nearly all of the non-associated gas was still shut-in. With estimated average production costs of $3 to $4 per mmbtu, Ouachita gas did not justify pipeline connections as long as federal regulation saddled the wells with conventional price ceilings. The thrust gas, on the average, was neither deep enough nor tight enough to qualify for higher or exempt pricing.

## Exploration in "Deep" Sedimentary Basins

The NGPA did, however, unleash a wave of exploration activity in the Anadarko Basin of western Oklahoma and the Texas Panhandle, where rich gas deposits could be found deep enough to qualify for exemptions from wellhead price controls. (See Map 6–4.) Deep gas discoveries also sparked interest in the Williston Basin of North Dakota and eastern Montana. Despite thirty years of oil and gas production in the basin, wildcat drilling probed strata that was deeper and acreage that was distant from historically prolific areas. Finally in March 1982, completion of the Ozark Gas Transmission System opened up a market for new gas discoveries in the Arkoma Basin of Arkansas.

Renewed interest in our nation's sedimentary basins is, in part, a product of recent scientific insights. Geophysicists at Cornell University reported in 1982 that while some basins are relatively shallow (the Illinois and Michigan basins are thought to contain only about

**Map 6-4.**   Deep Gas Wells.

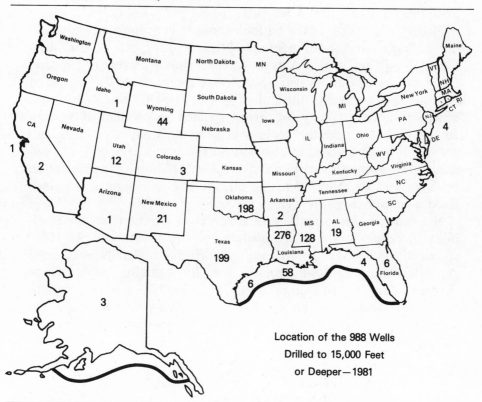

Location of the 988 Wells
Drilled to 15,000 Feet
or Deeper—1981

Source: Petroleum Information Corp, *Resume 81: The Complete Yearly Review of U.S. Oil and Gas Activity*, 1982.

12,000 feet of sedimentary deposits on top of the bedrock), others (like the Anadarko and Williston) may contain up to 30,000 feet of sediments below what has long been considered the bedrock basement.

The recent emphasis on deeper drilling in overthrusts and basins does not necessarily mean that these horizons hold the greatest resource potential. But deep gas certainly appeared to be the most profitable exploratory target prior to the 1985 broadening of deregulation scheduled in the Natural Gas Policy Act. The nation may be surprised to discover in the latter part of the decade that a great deal of gas exists in the midrange strata above 15,000 feet and as extensions of previously discovered fields and reservoirs. Such gas would be cheaper to produce than deep gas, but until deregulation it may be economically unavailable. For example, it is widely believed in Kansas that tremendous gas potential may exist in untested strata below currently producing depths in the Hugoton area, where producing wells now penetrate to only about 3,500 feet.

Finally, discussion of the outlook for U.S. conventional gas would not be complete without mention of frontier prospects on the Outer Continental Shelf and in Alaska. In 1981, only 1 percent of the total OCS acreage was under petroleum lease, compared to 21 percent of the total land surface of the United States. The OCS and Alaska are, however, more attractive for their suspected endowment of oil than for gas—not because the geology is necessarily expected to be oil-prone but because so many easy gas prospects still exist on the mainland. Moreover, when drilling takes place offshore, gas production becomes costlier than producing oil, for gas development demands fixed pipelines from the well to the mainland.

Not only is Alaska's offshore area truly a frontier, but the feasibility of developing even its onshore reserves is in doubt. In 1982, sponsors of the Alaska Highway gas pipeline officially postponed groundbreaking for two years—in reality signaling an indefinite delay. With a construction price tag in the $40 billion range, high interest rates, and the nation awash in gas, the ten-year-old concept for shipment of 26 tcf of proved North Slope reserves was either too late or too far ahead of its time.

## THE OUTLOOK FOR DOMESTIC SUPPLIES OF UNCONVENTIONAL GAS

The definition of *unconventional* gas is really a moving target. Gas is considered unconventional only because technology is still too primitive to justify its development at prevailing market prices. Deep gas in the Anadarko Basin, for example, was unconventional and virtually untouched until 1978, although shallower horizons in the basin had been thoroughly explored during the previous fifty years. Deregulated in 1978, the deep Anadarko became a very attractive prospect for exploration and development drilling.

Enormous volumes of unconventional gas are known to exist. These resources are usually grouped into six categories: deep gas, tight gas, gas from Devonian shales, gas from coal seams, geopressurized brines, and Arctic and subsea hydrates.

The first category, *deep gas* (including deep horizons in both sedimentary basins and overthrust belts), has already been discussed because much of it has now become conventional. (See Map 6–3.) Still, encouraged by a brief spate of unrealistic transactions as high as $10 per mmbtu, much of the post-NGPA drilling activity may have naïvely been aimed at truly unconventional depths. Of the six kinds of unconventional gas, an estimate of the ultimate resource boundaries for deep gas is perhaps the haziest. The industry simply is not drilling much beyond currently economical horizons.

To the contrary, petroleum companies encounter the other kinds of unconventional resources during their routine drilling activities. *Tight gas*, for example, is abundant in relatively shallow horizons throughout the Rocky Mountain states. It is particularly widespread in the San Juan Basin, sandwiched between more permeable rocks currently under production.

The NGPA has already brought the better horizons of tight gas into the marketable realm. As of early 1983, producers in sixteen states had managed to have their wells designated as "tight sands." (See Map 6–3.) 1982 production from tight sands accounted for about 5 percent of total domestic gas sales. In addition, if innovations in fracturing techniques can bolster well productivity, more and more tight gas may cross the threshold of conventional supplies.

In a 1980 report entitled *Unconventional Gas Sources*, the National Petroleum Council estimated that from 300 to 500 tcf of tight gas would become economical at wellhead prices of about $5 per mmbtu (1980 dollars). Ultimately, 900 tcf might exist in the sedimentary basins scattered across the United States. That volume would satisfy our nation's present gas habit for forty-five years.

The potential resource trapped within tight *Devonian shales* is even greater than the largely-western tight sands. Muds deposited 350 million years ago can now be found as shales, hidden at various depths beneath a fourth of the North American continent. Shales are perhaps richest and most accessible in the Midwest and Appalachian regions, for which the National Petroleum Council projected reserves in excess of 2,000 tcf—enough for a hundred years of U.S. consumption.

Within the same region, *gas from coal seams* might contribute another twenty years' worth of reserves. Coal-mining operations in Pennsylvania have been selling gas from coal beds for many years, while an estimated 250 mmcf is vented into the atmosphere each day. Map 6–5 illustrates the major coal basins in the United States.

The National Petroleum Council did not attempt to guess at the total resource of gas trapped in *geopressurized brines* beneath the Gulf Coast states. Producing horizons may run as deep as 50,000 feet. The Gas Research Institute in 1979 projected that the Gulf Coast area alone might ultimately hold between 3,000 and 50,000 tcf. That translates

**Map 6-5.**    U.S. Coal Basins.

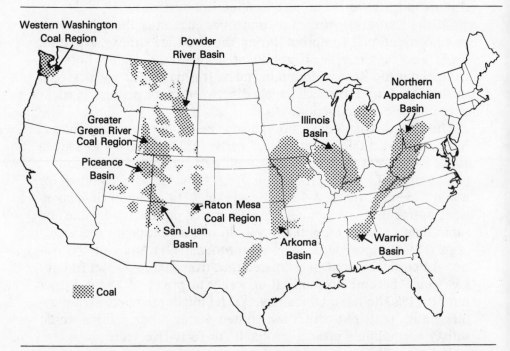

Source: American Gas Association, *Gas Energy Review* (September 1982) (adapted from Bryer, Malone, and Hunt—1982).

into 150 to 2,500 years worth of stock at present consumption rates. Geopressurized brines, however, present awesome environmental obstacles. No one has yet come up with a scheme for disposing of the hot and often-corrosive saltwater.

The 1980 report is too outdated to shed any light on the latest category of unconventional gas to receive serious industry attention. Located just barely beneath ground level in the earth's coldest regions, Arctic permafrost and deep sea sediments may contain massive amounts of methane trapped in crystals of ice. Recent evidence gathered by the *Glomar Challenger*, operating off Guatemala, suggests that subsea hydrates exist even in tropical waters—and at depths as shallow as 1,300 feet. Unleashing the energy within these *methane hydrates* is an attractive objective for technological research.

Finally, a discussion of unconventional gas resources would be incomplete without mention of *synthetic* gas opportunities. Especially with respect to recycling and use of our nation's sludges, wood chips, and other biological wastes, techniques for manufacturing methane are indeed worthwhile pursuits. Barry Commoner, perhaps the most famous of all critics of Big Oil, has prophesied methane would be the transitional fuel toward a solar-energy future precisely because it can be made from just about anything organic (Commoner 1979:56).

Coal gasification (as discussed in Chapter 2)—though synthetic rather than unconventional, by definition—is another frontier area for technological research. During the Carter administration research in coal gasification technologies was heavily subsidized. That substantial investment may eventually pay off, but the plight of the Great Plains Coal Gasification plant (under construction in 1983) casts doubt upon the commercial viability of any such venture today. (See Chapter 4.)

The establishment of research and development priorities and subsidies and incentives by the federal government is probably not the most efficient way to encourage development of abundant, economical, and environmentally benign energy sources. The experts were usually wrong in the past about which frontiers were most promising, and they are likely to be wrong today.

If the United States had had a Department of Energy and a national energy policy sixty years ago, we would have undertaken a monstrously wasteful campaign to drill wells on a grid pattern all over the public lands and a crash program to develop synthetic fuels. We might well have subsidized these projects by taxes, entitlements, or *rolled-in pricing* with respect to conventional crude oil. Very likely, we would have also subjected conventional fuels to price controls, windfall profits taxes, and administrative allocation because, after all, the experts believed we were liquidating the last of a disappearing resource. These burdens, along with the channeling of research toward alternate energy forms, might well have stifled or delayed the seismic revolution that did occur.

Today, the government-supported research projects that compare with grid drilling on the public lands are such undertakings as Lurgi-method coal gasification, Fischer-Tropsch coal liquefaction, and the mining of oil shale and tar sands in order to haul them to retorts and to the pit again. Each of these methods would produce fuels costing far more than oil prices at the peak of OPEC's power in 1981—even by their sponsors' most optimistic estimates. The governmental advocates of this "first generation of commercial production" do not justify synfuels as a cost-effective way of replacing OPEC oil but as a "widening of options" for the distant future.

This justification, however, rests on weak foundation. Each of these projects incorporates dead-end technologies with huge and inescapable burdens in handling of ore and spent rock and in consuming water. None of them has any significant opportunity for cost-saving breakthroughs, and each of them promises a host of adverse engineering and environmental surprises.

Industry would probably ignore all of these synthetic-fuel technologies if it were not offered some form of direct subsidy by the government or some governmentally enforced consumer subsidy. Each of these enthusiasms is an example of the orderly, linear bureaucratic reasoning that starts from some point on the present frontier of resource accessibility and marches straight up the steepest cost gradient in sight.

Despite the subsidies and in light of a turnaround in world petroleum supplies and pricing, most of the big synthetic oil and gas projects launched in the 1970s in both the United States and Canada have failed. By 1983, the central energy planning of the Carter administration was defunct, and the Synthetic Fuels Corporation was hanging on by a thread. But the nation was still a long way from adopting a philosophy of regulation that would "let a hundred flowers bloom" and from encouraging every branch of the industry to look for new approaches to energy supply by letting private enterprise face both the risks of devastating failure and the rewards of great success.

Today's leaders could do well by taking the Carter approach and standing it on its head. Instead of telling Americans that they must sacrifice to stave off an energy crisis and that nobody will be allowed to reap windfall profits from the effort, the president could proclaim that the OPEC price upheaval created opportunities for American individuals and corporations to become fabulously rich—by inventing and implementing new means to produce and conserve energy while serving the nation and the world at large. Perhaps no one would believe such an invitation, however, and perhaps no one should. The history of energy regulation and taxation in the United States has been so unstable that anything proclaimed or promised today is likely to be forgotten or ignored tomorrow.

## THE ROLE OF CONSERVATION IN OUR NATION'S ENERGY FUTURE

"Energy conservation" is a label applied to a broad spectrum of human activities: improved oil- and gas-production field practices; investment in more energy-efficient combustion devices; and changes in equipment or in use patterns so that less energy is required to achieve the same result (e.g., installing insulation). Conservation can also connote hardships if energy-intensive industrial plants are closed and workers are laid off and if low-income citizens are induced or compelled to go without heat or to turn down their thermostats to uncomfortable and sometimes life-threatening levels.

Because conservation has the same net effect on the national energy balance as an equivalent increase in energy output, some states (most notably California) have included "conservation" as a component of production forecasts in their planning documents. Recent history does, in fact, justify the special attention devoted to conservation. Between 1978 and 1981, the average residential gas customer in the United States shaved annual gas purchases by 14 percent. The 1981 average of 104.4 million btu per household represented the lowest rate of consumption since 1959.

One development that could cause a dramatic change in the gas-consumption outlook hinges on technological improvements in *fuel cells*. Fuel cells convert natural gas and other gaseous fuels into electricity and heat via an electrochemical process that avoids the inefficiencies of ordinary combustion. Moreover, by emitting both heat and electricity, they can fulfill virtually every facet of energy demand in homes and small businesses. Scientists are searching for design break-throughs that would make fuel cells an affordable investment for multifamily dwellings and commercial buildings. The National Aero-nautics and Space Administration, which pioneered early fuel-cell technologies, was in 1982 managing a fuel-cell test program for the Department of Energy and the Gas Research Institute.

It may be a long time, however, before fuel cells are widely mar-keted. Moreover, while widespread adoption of fuel cells would cer-tainly bolster conservation from the standpoint of total energy use, it is not clear whether that would result in a net increase or decrease in gas demand. Electricity generated in gas-fired fuel cells would dis-place central-station power generated by coal, oil, and gas, but the attractive thermal efficiencies of the cells could overwhelm this fuel-substitution effect.

Two other examples of conservation technologies are already suit-able for household use. Installation of *"pulse" gas furnaces* can increase combustion efficiencies from current levels of around 55 percent to almost 80 percent. The *electric heat pump,* too, has con-tributed to gas conservation in some regions of the country.

The electric heat pump furnishes heat neither by fuels combustion nor by electrical resistance. It simply extracts heat from the out-of-doors and channels it indoors. The remarkable thing is that a heat pump is effective even when outdoor temperatures are far colder than ambient indoor temperatures. Heat pumps can function in the opposite direction too—extracting heat from inside a building and moving it out in exactly the same fashion as electric refrigerators have operated for decades.

Although initial installation costs are higher than for conventional space-heating and cooling equipment, heat pumps are an attractive investment because their thermal efficiencies tend to be far greater than the alternatives, and *life-cycle costs* are therefore lower. Thermal efficiencies are even more pronounced for heat pump designs that draw heat from subsurface soils and waters (of relatively constant year-round temperatures) instead of the atmosphere.

The growing popularity of electric heat pumps will thus tend to de-press gas consumption in two ways. First, households and businesses are choosing the new electrically driven technology in place of gas heating. Second, the superior cooling efficiency of heat pumps re-duces the hot-weather demand for electricity and thereby the need for gas as a fuel for electric-utility peaking plants.

The conservation opportunities offered by all of these technical achievements can be bolstered by the efforts of state regulators. For example, because electricity prices have long been subject to crude-oil price increases, state public utility commissions have instituted incentives to encourage electric utilities to promote conservation efforts by their customers. Foremost are policies that allow electric utilities to include conservation gains in company rate bases. Profits can thereby be earned from conservation as well as the installation of new capacity. Given the price run-ups for natural gas in the early 1980s, it would appear that state conservation commissions would be wise to extend this incentive to gas-distribution companies. But the high prices themselves have already induced so much conservation and fuel-switching that the resultant loss in load has reduced revenues for many companies to dangerous levels.

## THE OUTLOOK FOR GAS IMPORTS

Although this text intentionally concentrates on natural gas in the United States, one cannot consider the supply outlook without taking a wider view. Projecting future supplies of gas from abroad takes even more imagination than confronting our own geologic, technologic, and economic unknowns. With respect to gas imports, forecasters are also faced with the need to make political conjectures that can turn geologic and economic logic upside down.

### International Gas Resources

Official tabulations of known gas reserves worldwide are likely to be even more misleading than those published for the United States. While regulation has discouraged gas exploration in the United States, a dearth of interested purchasers has effectively stifled gas exploration in all but a few industrialized nations in the Northern Hemisphere. Practically all of the gas reserves now known to exist abroad owe their discovery to the quest for oil. Why, after all, should anybody look for gas when in 1980 fully half of all associated gas produced in conjunction with oil (and not reinjected) was flared at the wellhead? (See Chapter 4, Table 4-1.) Why would a nation bother to develop its nonassociated gas reserves when it is still unable to market all of the associated gas that its oil production program brings forth?

From a world perspective, gas is truly the black sheep of the hydrocarbon family. It is one of the easiest fuels to burn, and certainly the cleanest, but it is a bother to move from one place to another. Land-based movement is not so difficult; fixed pipelines are often the preferred vehicle for transporting even crude oil and refined products. Unfortunately, most of the world's energy markets are oceans away from producing countries, and tanker transport for methane in the supercooled form of liquefied natural gas (LNG) requires docking and shipping equipment that is an order of magnitude more sophisticated (and more costly) than similar facilities for crude oil. Half of all crude oil produced, for example, is exchanged on the world market. Only a tenth of all commercial gas production crosses international borders, and 90 percent of those transactions involve fixed pipelines.

For all of these reasons, reserve statistics shed little light on the ultimate extent of international gas resources. Perhaps the best indication is a comparison of U.S. and worldwide sedimentary basins. Almost 90 percent of these hydrocarbon-prone environments lie outside of U.S. jurisdiction. Whatever we may find in the United States, there is a good chance that the rest of the world supports nine times as much.

## Prospects for International LNG Trade

If international exchange of natural gas is to increase, it will have to rely on LNG technology. Interest in LNG (in the United States and to a lesser extent in the other major gas-importing countries of Western Europe and Japan) peaked in the 1970s, with the 1980s bringing a downturn. A surplus of world oil supplies, coupled with impressive progress in energy conservation (only partly attributable to the worldwide recession), has dampened LNG enthusiasm among the prospective gas-importing nations. Ironically, this comes at a time when interest among potential gas exporters has increased. Canada and Australia, for example, were both attempting to put together international LNG sales in 1983.

The prime obstacle to increased LNG trade, however, has been the failure of political leaders in oil-exporting countries to face up to changing economic times. As long as producing nations insist that gas should command an export price on a par with crude oil, only fools (and desperate ones at that) would buy it. LNG that is loaded, for instance, in Algeria at a price equivalent to crude oil will cost a lot more than its btu-equivalent of refined product by the time charges for cryogenic tankering, insulated storage, regasification, and downstream transmission and distribution are loaded on. Since few uses exist for gas that cannot be accomplished just as well by oil, why not opt for the cheaper fuel?

During the 1970s, however, when the crude-oil shortage seemed both real and entrenched, Japan was desperate, Western Europe was desperate, and the United States was desperate. They all signed purchase agreements, primarily with Algeria and Indonesia, and embarked on costly shipbuilding and terminal ventures.

In 1981, Japan received half of the 1.6 tcf of methane traded internationally as LNG. Despite construction of three major LNG receiving terminals, however, the history of LNG imports into the United States has been an utter failure. The biggest U.S. project was shut down within a year of initial deliveries because the government refused to acquiesce in Algeria's unilateral decision to change the price terms, and when a second venture came on line in September 1982, distributors and industries facing huge rate increases protested continued shipments. (See Chapter 4 for a detailed discussion.) In 1981, the United States purchased only 0.04 tcf of foreign LNG, which was less than what it exported from Alaska to Japan that same year.

At least for the near term, expansion of the LNG business may not be in the best interest of either importing or exporting nations, even if the alternative is to continue or increase gas flaring. Like money, energy is fungible. The sale of one more unit of gas generally means the sale of one less unit of oil. As long as oil is cheaper to produce and ship than gas (even assuming a zero wellhead cost for gas), sheer efficiency would favor production of oil, and producing countries as a group would reap greater profits.

Ultimately, the prospect for increased LNG trade may become the most important consideration within the whole spectrum of natural gas issues facing industrialized nations. But in the early 1980s, amidst a crude-oil surplus and a downturn in energy consumption, it is certainly not as pressing as it appeared in the 1970s.

## The Outlook for Gas Imports from Mexico

The United States, more than any other energy importer, can afford to ignore the ups and downs of the LNG business. Our nation has the good fortune to border two countries whose supplies of conventional gas appear to dwarf their own internal energy needs. Tapping these vast Canadian and Mexican reserves via pipeline connections is technologically and economically feasible. In fact, the United States has been importing gas from Alberta since the late 1950s and from Mexico's richest hydrocarbon province since early 1980. (See Table 6-2.) 1981 imports from Canada totalled 762 bcf at an average price of $4.83 per mmbtu, and Mexico supplied 105 bcf at a price of $5.01.

**Table 6-2.** U.S. Gas Imports and Exports.

| | Exports (millions of cubic feet) | | |
| --- | --- | --- | --- |
| Year | Total | Canada | Mexico | Japan[a] |
| 1955 | 31,397 | 11,494 | 19,903 | 0 |
| 1960 | 16,100 | 5,574 | 10,526 | 0 |
| 1965 | 27,428 | 17,892 | 9,536 | 0 |
| 1966 | 54,860 | 44,958 | 9,902 | 0 |
| 1967 | 81,595 | 70,456 | 11,139 | 0 |
| 1968 | 93,745 | 81,647 | 12,098 | 0 |
| 1969 | 51,308 | 34,936 | 13,390 | 2,982 |
| 1970 | 69,813 | 10,860 | 14,678 | 44,275 |
| 1971 | 80,365 | 14,349 | 15,785 | 50,231 |
| 1972 | 78,014 | 15,553 | 14,579 | 47,882 |
| 1973 | 77,169 | 14,824 | 13,999 | 48,346 |
| 1974 | 76,789 | 13,263 | 13,268 | 50,258 |
| 1975 | 72,675 | 10,219 | 9,454 | 53,002 |
| 1976 | 64,710 | 7,506 | 7,425 | 49,779 |
| 1977 | 55,626 | 31 | 3,940 | 51,655 |
| 1978 | 52,533 | 66 | 4,033 | 48,434 |
| 1979 | 55,673 | 76 | 4,308 | 51,289 |
| 1980 | 48,731 | 113 | 3,886 | 44,732 |
| 1981 | 59,402 | 106 | 3,367 | 55,929 |

Source: American Gas Association, *Gas Facts: 1981* (Arlington, Va.: American Gas Association, 1982). Adapted from U.S. Department of Energy, Natural Gas Division.

[a]LNG exports from Alaska.

## Table 6-2 (continued).

### Imports (millions of cubic feet)

| Year | Total | Canada | Mexico | Algeria[a] |
|------|-------|--------|--------|------------|
| 1955 | 10,892 | 10,885 | 7 | 0 |
| 1960 | 156,843 | 109,855 | 46,988 | 0 |
| 1965 | 456,694 | 404,687 | 52,007 | 0 |
| 1966 | 480,591 | 431,955 | 48,636 | 0 |
| 1967 | 564,228 | 513,256 | 50,972 | 0 |
| 1968 | 651,885 | 604,462 | 47,423 | 0 |
| 1969 | 726,952 | 680,107 | 46,845 | 0 |
| 1970 | 820,781 | 778,688 | 41,336 | 757 |
| 1971 | 934,547 | 910,925 | 20,689 | 2,933 |
| 1972 | 1,019,495 | 1,009,092 | 8,140 | 2,262 |
| 1973 | 1,032,903 | 1,027,216 | 1,632 | 4,055 |
| 1974 | 959,285 | 959,063 | 222 | 0 |
| 1975 | 953,007 | 948,114 | 0 | 4,893 |
| 1976 | 963,768 | 953,613 | 0 | 10,155 |
| 1977 | 1,010,431 | 996,723 | 2,384 | 11,324 |
| 1978 | 965,545 | 881,123 | 0 | 84,422 |
| 1979 | 1,253,383 | 1,000,775 | 0 | 252,608 |
| 1980 | 984,767[c] | 796,507[c] | 102,410 | 85,850[d] |
| 1981 | 903,950 | 762,113[b] | 105,013 | 36,824 |

Source: American Gas Association, Gas Facts: 1981 (Arlington, Va.: American Gas Association, 1982). Adapted from U.S. Department of Energy, Natural Gas Division.

[a]Quantities represent total LNG imports from overseas into the United States.

[b]Includes 6 million cubic feet of LNG.

[c]Volume does not include 282,112 mmcf of intransit natural gas receipts.

[d]Volume delivered by state; Massachusetts—24,274 mmcf, Georgia—24,603 mmcf, Maryland—36,973 mmcf.

Together, these continental imports accounted for almost 5 percent of U.S. gas consumption.

Politics, however, can play havoc with even the soundest economic and engineering principles. U.S. experience with Canadian and Mexican gas imports in recent years has not been much smoother than its experiences with Algerian LNG. Sometimes the political obstruction is inherent in the host country, as was the case during the mid-1970s when Canada was just as preoccupied with its own energy security as was the United States. Sometimes, however, our nation's inability to secure continental imports reflects our own insensitivities and diplomatic naivety.

The sterling example of U.S. ineptitude was the U.S./Mexican gas negotiations that took place during President Carter's term in office. Shipments from the Reynosa Field just south of the border had dwindled from a peak of 52 bcf in 1965, and in 1975 they ended altogether. Meanwhile, vast amounts of associated gas in the southeastern Reforma Field were flared, pending development of markets and a transmission system.

During negotiations in the mid-1970s, Petroleòs Mexicanos (Pemex, Mexico's national oil company) was seeking border-price parity with distillate fuel oil; the U.S. negotiating team (headed by Secretary Schlesinger of the fledgling Department of Energy) advocated parity with lower-cost residual oil. In hindsight, Secretary Schlesinger was correct in arguing that natural gas ultimately competes with residual oil, not distillate fuel. (See Chapter 7 for a discussion of gas marketability.) But given the conditions of the time, he was wrong.

U.S. regulation of domestic prices had created a supply shortfall that had to be filled by something. Compared with deep gas, synthetic gas from coal, OPEC LNG, or gas from the Arctic, Mexican gas at heating-oil parity was a bargain. Very few observers envisioned that marketability problems for this "prince of fuels" could ever arise.

Mexico was adamant. Why should it settle for a price less than that of distillate oil when countries like Algeria and Indonesia had negotiated contracts for LNG exports that were expected to yield prices (landed in the United States) two or three times that amount? Perhaps even more important to Mexican leaders than the reality of striking a fair deal was the need to ensure that within its own country the transaction appeared to be reasonable. Mexico had, after all, suffered the usual discomforts of multinational exploitation prior to nationalizing its petroleum holdings in 1938. After having worked so hard to expel U.S. companies in the oil patch, it would be unthinkable for the government to now acquiesce to a more subtle, but nonetheless potent, form of subjugation.

Overall, the situation called for extraordinary finesse. The United States, however, approached the negotiating table like a bull in a china shop. Our negotiators knew that Mexico had to sell its gas. The country was determined to increase its oil exports and thereby earn revenues sufficient to support an ambitious program of social and economic development. Increased oil production meant increased production of associated gas, very little of which made sense to reinject. Absent an export market, that gas would have to be flared. With so many other oil-producing nations facing the same problem, the infant LNG trade did not offer Mexico much hope.

Realistically, the United States was the only buyer for the enormous volumes of gas Mexico anticipated producing. Even accounting for huge increases in its own industrial gas consumption, Mexico hoped to export about 2 bcf per day, expanding to 4 bcf within a few years. That amount of gas was significant: Two bcf per day was what the Alaska gas pipeline sponsors promised to deliver. Surely it would be in Mexico's interest to sell the gas even at the residual-oil price, since OPEC had ensured that the going price for any oil product already incorporated enormous windfall profits. The United States, in fact, had reason to believe that its price offer was more than fair; producers of gas within its own borders were still subject to wellhead price regulation that held prices far below that of residual oil.

Despite the apparent impasse, Mexico made plans to begin construction of a 48-inch trunkline that ultimately would stretch 750 miles from the Reforma Field to the Texas border, with spurs to Mexico's own industrial regions. The Export-Import Bank of the United States (Exim) tentatively agreed to loan Mexico $590 million for the project. In response, the chairman of the Senate Subcommittee on International Finance (whose jurisdiction included oversight of Exim activities) introduced a resolution to halt the transaction pending Mexico's commitment to a gas export contract at a price agreeable to the U.S. Department of Energy.

Senator Stevenson's bill did not reflect the majority mood of Congress; it never even reached the floor for a vote. But Mexican leaders were incensed, and Mexico's president announced that his country would not renew the Memorandum of Understanding between Pemex and a half dozen U.S. pipelines, which was due to expire at the end of the year (1977). Soon thereafter, Mexico was able to put together pipeline-construction capital loaned by European banks.

Determined to use all the gas within Mexico now, Pemex commenced construction of the transmission system and instituted a program for converting Mexican industries from oil to gas. In March 1979, the 48-inch pipeline construction halted just short of the U.S. border, at a connection with an existing transmission system that had carried gas from the Reynosa Fields to industries in the north-central part of the country.

Two events occurred in June 1979 that changed the bargaining positions of the two countries. Oil prices increased 25 percent, and a Mexican oil well in Gulf waters (Itox 1) blew out. The blow-out eventually earned the status of being the worst oil disaster anywhere in the world; Mexico lost an estimated 3.1 million barrels of oil and 3 bcf of gas before the well was capped in March 1980. Moreover, the incident was a severe diplomatic embarrassment, since some of the oil washed up on Texas beaches.

In September 1979, the two countries agreed upon the terms for a gas sale. By then, however, Mexico had made remarkable progress in attempting to use most of its surplus gas internally—an event nobody in the United States had really anticipated. As a result, Mexico was no longer willing to sell 2 billion cf per day; the agreement provided for only 300 million cf per day. The United States, however, was successful in getting a lower price than Mexico had initially stated. The terms of sale called for a price that was the weighted average of residual oil and No. 2 fuel oil, based on an 80/20 split. While U.S. companies retained the right to cancel or reduce their takes (with six-months' notice), Mexico had the unilateral right to adjust its volume offerings below the 300 mmcf ceiling.

In some respects, the outlook for increased Mexican gas in the 1980s looks pretty good. Although official reserves data would not show it, it is widely viewed within the industry that Mexico's ultimate hydrocarbons potential may be second only to Saudi Arabia's.

In 1975, virtually all gas supply forecasts projected that by 1980 imports from Mexico would tally 2 or even 4 bcf per day. There is no reason to believe that forecasts made today will be any less wrong. Nevertheless, Mexico's near-default in loan repayments, which shook the international financial community in mid-1982, does lend credence to speculation that gas exports may increase.

Like many countries newly awash in petrodollars, Mexico decided to use its money to create more money. But it went too far. Overleveraged and caught unprepared by the 1982 down-turn in world oil prices, Mexico could not meet its scheduled payments. The solution? Perhaps Mexico will increase its crude-oil production, with consequent increases in natural gas availability.

It must be remembered, however, that Mexico can flare a great deal more gas without approaching the scale of flaring in Africa and the Middle East. In 1980, Mexico flared only about 16 percent of its total gas production. This compared to an average flaring ratio of 48 percent in OPEC countries, with Saudi Arabia flaring 72 percent and Nigeria wasting 96 percent of its resource. (See Chapter 4, Table 4-1, for details.)

One other difficulty for the United States in dealing with Mexico is the fact that transactions cannot be viewed in isolation. For example, when the 1979 deal with Mexico was cut, Canada reacted to the fact that its own border price was about $1 per mmbtu less than what Mexico had been able to extract. It began a series of increases, keeping pace with escalating oil prices. By the spring of 1981, the Canadian price was about ten cents higher than the price Mexico was getting under its contract formula. Not surprisingly, Mexico then demanded and received price parity with Canada.

## The Outlook for Gas Imports from Canada

Although the history of U.S. trade and diplomatic relations with Canada has been much better than our record with Mexico, the Canadian gas business is in such turmoil that forecasts of future export prospects are perhaps even more in doubt. If the U.S. energy industry believes that government energy policy has played havoc within the United States, its leaders can console themselves by reflecting on the plight of their northern counterparts.

Canada's biggest problem in the early 1980s is that it has not been able to shake the outlook it held in the 1970s. In the last decade, conventional wisdom held that *nonrenewable energy resources* were both finite and near exhaustion. Although the country was exporting about half of its annual production under long-term contracts negotiated before the energy crisis struck, it was politically unacceptable to sanction any new export licenses unless either the gas was simply "loaned" to the United States (to be replenished when the Alaska gas pipeline was complete) or else it came from the Canadian Arctic. (By contrast, Mexico was exporting less than a tenth of its production even in 1980 when Reforma gas began to flow across the border.) And for whatever volume of gas Canada relinquished to the United States, there was little doubt that Canada would demand (and the United States would accept) a price that was commensurate with its importance to both nations.

Despite the changed circumstances—from a U.S. gas shortage to a gas glut—the viewpoint born of the 1970s lingers. Superimposed on that philosophy is a gas supply situation that calls for just the opposite approach. In 1982, Canada had about 70 tcf of known reserves (90 tcf if Arctic reserves are included). This was enough to satisfy 1982 levels of domestic consumption for thirty-five to forty-five years, and nobody questioned that a huge volume of gas was sitting just outside of the "proved" category (most notably in the Deep Basin of Alberta), pending economic conditions that would make it fruitful to invest in field delineation wells. Moreover, by the fall of 1981, an estimated 14,000 gas wells were *shut-in* for lack of a market.

Producers and pipelines pleaded with the National Energy Board (Canada's equivalent to the U.S. Federal Energy Commission and the Economic Regulatory Administration combined) to open up new export opportunities. But by the time the National Energy Board decided to act on the industry's proposals, U.S. buyers found that Canadian gas was a commodity they could no longer afford—at least for the short-term.

At a $4.94-per-mmbtu *border* price, U.S. pipelines and their distributors found that Canadian gas (in addition to costly deep gas, tight gas, and imported LNG) made their systemwide supplies too expensive. Industrial customers began switching to other fuels, and the remaining commercial and residential customers faced huge rate increases because the fixed costs of transportation had to be covered by fewer buyers. By late 1982, U.S. companies were taking only about half of the Canadian gas available under existing export licenses, and one traditional buyer (Natural Gas Pipeline Company of America) was taking none at all.

By then, Canadian producers (and the western provinces dependent on the taxes and royalties from oil and gas sales) were in a cash crunch. On the one hand, it seemed prudent to attempt to bolster sales by cutting the price. On the other hand, why should Canada cut its prices if deep gas producers in the United States would not? Moreover, with an estimated 2 tcf of annual surplus deliverability from domestic wells in the U.S., price-cutting on Canada's less than 1 tcf of sales looked like a bottomless pit.

An additional obstacle is the fact that Canada is a dominion of strong provinces with very different needs and desires. Provincial/federal negotiations over oil and gas export prices, export volumes, and the splitting of taxes, is always a high-stakes undertaking in which much extraneous trading stock is brought to the table. The 1980 negotiations, for example, were far from congenial. Separatist factions were powerful not only in Quebec but also in the western provinces, where business and government leaders openly scorned the eastern provinces as welfare states grasping for ever-increasing chunks of the western petrowealth.

Canada, in essence, cannot tamper with the border price of gas sold to the United States without upsetting a complex of painfully arrived at political compromises. And past investments and political capital it has spent on its drive to find and develop gas in the Canadian Arctic and to bring gas to the eastern provinces cannot be overlooked.

In the 1970s, the national oil company, PetroCanada, directed much of its endowment of public funds toward Arctic exploration. A consortium of companies operating as PanArctic Oil in the Arctic Islands, along with Dome Petroleum drilling in the Mackenzie Delta to the west, also invested large amounts of money in exploration. As much as 90 percent or more of that money would have been paid to the national treasury as corporate taxes had not the special Arctic tax credit been available.

Both the western Arctic around the Mackenzie delta and the eastern Arctic Islands proved to be poor in oil but rich in gas. In order to profit from any part of the investment, costly pipelines or LNG facilities were inevitable. Transportation schemes blossomed, and then shifted and recombined. The Mapleleaf Project, Polar Gas, and Arctic Pilot LNG were the biggest. Of them, the $4 billion Arctic Pilot project is the only one that was still appearing in the trade press in 1982. Recognizing the growing disenchantment for LNG imports in the United States, the two partners (TransCanada Pipelines and PetroCanada) ceased negotiations with Tenneco and courted German buyers instead. In late 1982, however, Europe's insistence upon a delivered LNG price competitive with residual oil all but killed the project.

Perhaps the most awesome of gas-policy questions facing Canada is the dilemma of the eastern provinces. During the energy crises, the Maritime provinces as oil importers were vulnerable to world oil-price upheavals and supply insecurity. Imaginative programs were developed to spread the cost burden over the nation as a whole. But that was only tolerable as a stop-gap measure. Political leaders had to find a way to get the Maritimes "off oil." Eventually the Trans-Quebec and Maritime (TQ&M) project evolved, sponsored by TransCanada

Pipelines (Canada's only west-east gas carrier) and Nova Corp. (owner of the major intraprovincial pipelines in Alberta).

With construction underway in 1982 and 500 million of federal dollars invested by mid-1983, the TQ&M project faces formidable marketing obstacles. Alberta producers accepted a 30-cent wellhead-price reduction for all gas destined for the TQ&M system. The federal government and the local gas utility offered grants to entice eastern residential users to convert their furnaces. Meanwhile, both the refining industry and Quebec's big hydroelectric company prepared for battle.

Lured by the promise of a stable employment base, the eastern provinces had encouraged and even subsidized construction of huge refineries during the 1960s and 1970s. The capacity far outstripped export, not to mention local, demand. By the early 1980s, several plants had closed down, and others were operating at reduced capacities. Their biggest problem was finding buyers for the heavy residual oil that remains after the lighter hydrocarbons are distilled. Although several refineries planned joint construction of an upgrading facility to process the residual oil into more saleable products, unless the surplus in residual-oil capacity is in fact removed, refiners will always cut their prices for heavy fuels to whatever level is necessary to dispose of the entire product in the industrial boiler-fuel market.

That leaves only commercial and residential customers as prospective gas buyers, and these are being wooed by provincially owned utilities (Ontario Hydro and Hydro-Quebec) with major surpluses of electricity. Overcapacity is so great that in 1982 Hydro-Quebec, operator of the substantial facilities at St. James Bay, and New England utilities across the border worked on an export agreement for electricity.

Even if gas manages to capture a market sufficient to fill up the new TQ&M system, it is very likely that within a few years of completion, the pipeline will instead carry gas westward from the Sable Island area (off Nova Scotia), which has already turned up 3.5 tcf of known discoveries (though in deep sediments between 11,000 and 19,000 feet), with prospects for 12 tcf or more. That amount of gas would swamp eastern users, and it does not take much imagination to see that the TQ&M pipeline, heavily subsidized by Canadian taxpayers and Alberta producers, may be ideally suited instead for exporting gas to the United States.

The eastern dilemma approaches absurdity when one considers the recent oil discoveries in the Hibernia field off Newfoundland. Canada could rid itself of OPEC oil vulnerability without ever having to go "off-oil."

Despite the seemingly no-win character of the choices facing Canada in its Arctic exploration programs and in its energy initiatives in the eastern provinces, mid-1983 Canadian gas exports had fallen to such low levels that the government was forced to act. The border price was cut to $4.40 per mmbtu, and when that reduction failed to turn around the sales decline, Canada authorized special discounts. Under its Volume-Related Incentive Pricing program, gas was available on a short-term basis for as little as $3.40 per mmbtu. Meanwhile, U.S. officials were pondering the merits of pressing for a change in the uniform border-price policy so that individual companies might negotiate geographic differentials in contract prices. In 1983, policies on both sides of the border were changing so fast that it was impossible to predict for any particular action its long term ramifications with respect to the industry response in the United States or interprovincial politics in Canada.

## OUR INEXHAUSTIBLE GAS RESOURCES

Overall, the outlook for U.S. access to natural gas (both domestic and imported) is perhaps more difficult to translate into numbers than ever before. History tells us to be wary of economic forecasts and especially those that espouse Chicken Little's "the sky is falling" mentality; after all, it is the very prospect of a shortage (more accurately, increased prices) that drives exploration, conservation, and technological innovation. The nation's legacy of regulation suggests that U.S. companies have only begun to look for gas. The statistics of worldwide gas flaring indicate that it will be a long time before the rest of the world follows suit.

Geology tells us that much more gas than oil is likely to exist in the earth's crust, and we already know that vast methane resources reside just outside the bounds of economic feasibility, as defined by current technology. Economics tells us that we cannot talk intelligently about gas production and gas supplies without mentioning price and, in turn, without investigating how price affects gas demand among the various classes of consumers. While speculation about the adequacy and logistics of future gas supplies captivated industry and government attention during the seller's market of the 1970s, the 1980s may give birth to a whole new emphasis on gas marketability as the balance of power shifts from sellers to buyers. The following chapter attends to this issue.

## REFERENCES

Commoner, Barry. *The Politics of Energy*. New York: Alfred A. Knopf, 1979.

National Petroleum Council. *Unconventional Gas Sources*. Washington, D.C.: National Petroleum Council, December 1980.

Schanz, John J., and Joseph P. Riva, Jr. *Exploration for Oil and Gas in the United States: An Analysis of Trends and Opportunities*. Washington, D.C.: U.S. Government Printing Office, September 1982.

# 7 GAS DEMAND AND MARKETING PRINCIPLES

## A PREMIUM FUEL?

During the energy-crisis years of the 1970s, natural gas was acclaimed as "the prince of fuels." Its clean-burning qualities were indeed unsurpassed. Moreover, popular opinion held gas to be a fast-dwindling resource, and like wine of respectable vintage, its certain depletion promised an ever-increasing value.

The notion that gas was a premium fuel suggested, in turn, that gas customers would be willing to pay premium prices. Because this belief was virtually unchallenged for many years, it gave rise to a score of supplemental-gas plans mid-decade and to the late-1970s quest for costly deep-gas reserves. The premise was, nevertheless, fundamentally wrong. As prices rose, customers began to conserve to an extent that surprised almost all the experts and to flee to cheaper fuels. Gas supply began to outdistance demand by a wide margin, and a growing glut in the early 1980s threatened disaster upon those producers and utilities that had bet their financial futures on the conventional wisdom of the previous decade.

By 1983, the claim that gas was a premium fuel had almost disappeared from industry rhetoric. The role that natural gas would play in the nation's energy future no longer seemed obvious. Were the doomsayers correct, or had the gas "shortage" of the 1970s been

159

merely a shortage of producer incentives? Whether bearish or bullish, however, everybody in the industry now realized that the erosion of price controls initiated by the 1978 Natural Gas Policy Act meant that a company could no longer afford a complacent approach to gas marketing. No prudent marketing strategy could be based on the fanciful idea that gas was intrinsically superior to other forms of energy.

Just when the doctrine that gas was a premium fuel began to lose favor within the industry, natural gas was becoming more important to the nation as a whole. Unlike coal mining, the production of natural gas does not beget stripmined wastelands, nor does it foster black-lung disease. It does not sully beaches and birds with the products of its transportation mishaps. Natural gas does not threaten the wilderness and recreational values of free-flowing streams, nor does it impose the multigenerational responsibilities of nuclear waste.

Combustion of gas is relatively free of soot, carbon monoxide, and the nitrogen oxides associated with other fossil fuels (although like all fossil fuels, gas combustion does exacerbate the "greenhouse effect" of increased levels of atmospheric carbon dioxide). Sulphur dioxide emissions are almost nonexistent, and substitution of gas for high-sulphur coal and fuel oil could, therefore, provide the cure for North America's acid rain ills. Finally, natural gas produced within the United States offers hope for reducing the national trade imbalance occasioned by oil imports, and gas from anywhere in North America is beyond the control of OPEC nations.

Only through governmental design of incentive programs, prescriptions, and penalties will any of these social considerations influence what actually goes on in the market. From a public-interest standpoint, therefore, gas may indeed be a premium fuel, but except to meet certain air quality standards that already exist, the market simply does not view it as such.

## THE MYTH OF CAPTIVE MARKETS

Natural gas furnishes over one-fourth of the *primary energy* consumed in the United States, and more than half of the nation's non-transportation supplies of energy. The future may hold a place for natural gas even in transportation. Already, *compressed natural gas (CNG)* and *methanol* (methane that has been chemically altered into a liquid alcohol) have begun to penetrate markets in surface transportation. In Italy and New Zealand, particularly, compressed natural gas has captured a significant share of the transportation market. Technical and marketing breakthroughs may also allow *liquefied natural gas (LNG)* to displace far heavier kerosene as a fuel for jet aircraft.

Annual U.S. consumption of natural gas averaged about 20 trillion cubic feet (tcf) during the 1970s and into the early 1980s. The major uses of gas include space heating (and cooling), water heating, cooking in homes and businesses, industrial and electric-utility boiler fuel and turbine fuel, and as process gas for industries that use methane as chemical feedstock or rely on its superior flame characteristics for precision tasks. (See Figure 7–1.)

None of these uses creates a demand for natural gas as such. There is very little that gas can do that cannot also be achieved in some way by coal, oil, or electricity. Fuel substitution is easiest in boiler-fuel markets—electric utilities and industrial plants that can use a variety of primary energy sources to raise steam. *Interfuel competition* is sharp in many other *stationary-heating* applications, including plants that use the heat of combustion to assist in the physical or chemical transformation of materials. Oil refining, metals smelting, and cement manufacturing are prominent examples.

The choice of fuel by these large industrial consumers is, of course, affected by the flexibility of their existing equipment. For boilers

**Figure 7–1.**    Gas Markets by End-Use, 1982.

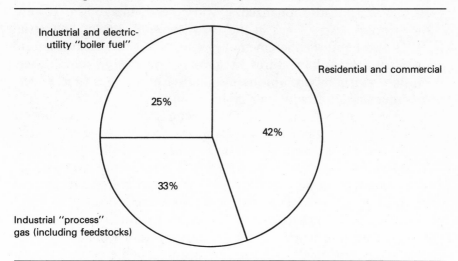

Industrial and electric-utility "boiler fuel"

Residential and commercial

25%

42%

33%

Industrial "process" gas (including feedstocks)

that are suited for gas and gas alone, suppliers of gas can normally extract some premium over the cost of competing fuels without risking load loss. Gas holds an additional advantage because it is the cleanest-burning of all fossil fuels. This advantage translates into cheaper plant maintenance and the avoided costs of pollution-abatement gear.

Pollution regulations differ markedly around the country, however. Industries in southern California, for example, are willing to pay more for gas than plants in sparsely populated (and coal-rich) areas of the mountain west. Indeed, in 1982, California was the only state where industrial rates for gas were higher than residential rates.

Despite the implied air-quality premium for natural gas, industrial customers in most parts of the country can (and do) abandon gas service at btu-equivalent prices barely above those of alternative fuels. This is because many boiler-fuel users of gas—especially those outside of the gas-exporting states—installed *dual-fuel capacity* (equipment capable of burning two or more kinds of fuel) when their supplies were curtailed in the 1970s. They also installed the requisite pollution hardware to curb emissions from high-sulphur coal or residual-oil combustion. In 1983, about half of the established industrial gas customers with large boilers were capable of switching to coal or oil, often at the flick of a switch.

Perhaps the biggest blunder of business and government leaders in the 1970s was their failure to recognize the importance of industrial customers to the systemwide operations of the gas industry. Congress, in fact, institutionalized a widespread antagonism toward "low-priority" uses of gas in the 1978 Power Plant and Industrial Fuels Use Act and the incremental-pricing provisions of the Natural Gas Policy Act. Both measures attempted to wean large-volume users from natural gas by prohibiting installation of new gas-fired boilers or by imposing discriminatory price penalties.

It is not surprising, therefore, that pipelines went about purchasing newly deregulated gas mainly with an eye to their "high-priority" residential and small commercial customers. Still feeling the sting of gas shortages and the attendant flood of customer suits over abrogation of supply contracts, transmission companies gobbled up deep gas (which the 1978 Congress had freed from price controls) at prices four or five times higher than the average of regulated gas prices.

Having spent the last forty years in a supply- rather than a demand-limited environment, the pipelines were almost oblivious to the pros-spect that the combination of increased supplies and higher prices might in fact clear the market. Worse yet, they failed to consider the risks and the devastating consequences if supply commitments went beyond market clearing—which indeed began to happen in early 1982. Virtually every interstate pipeline was caught unawares by the gas glut that swept the industry only four years after the Natural Gas Policy Act had changed the rules of the game.

As prices charged industrial gas users approached and then exceeded parity with residual oil, pipelines and distributors lost more than just their dual-fuel customers. Even the most "captive" gas users, like fertilizer plants dependent on methane feedstocks, opted out of their supply arrangements. They had no realistic alternative for feedstock, of course, but the high prices coupled with the severe recession were fatal. Finding it impossible to compete with Mexican plants having access to far cheaper supplies, domestic fertilizer plants in the Great Lakes region simply ceased operations.

While most pipelines and distributors by mid-1983 had come to recognize the dangers of industrial load loss (and were taking bold actions to prevent further losses and to woo back excustomers), there was little evidence that they had yet grasped the dangers of residential load loss. Indeed, in order to salvage their industrial markets, many utilities attempted to restructure their rates, shifting costs from their most price-sensitive to their least price-sensitive customers.

Forcing residential customers to bear the brunt of high gas-purchase costs may have seemed unfair, but few questioned the effectiveness of that approach for generating revenues sufficient to offset higher wellhead prices. Because the competition in the residential sector is No. 2 heating oil or even electricity, rather than the far cheaper No. 6 heavy fuel oil (or "resid") used by boiler-fuel operators, there was little prospect of fuel switching as long as residential rates remained below the No. 2 price. Fuel switching, however, is not the primary threat of utility load loss among residential energy users. The *price elasticity of demand* is a far more formidable hazard because it begins to cut deeply into sales volume at gas prices well below those of competing fuels.

While residential customers—at least over the short term—are captives of gas distributors because of the high front-end costs for replacing existing furnaces and because most of these customers have difficulty in appraising the nuances of long-term fuel savings, there is one thing that homeowners and tenants are remarkably good at: turning down the thermostat and layering windows with plastic. Moreover, price elasticity of demand—though it begins to make itself apparent almost immediately—tends to aggravate the load loss through time even if gas prices stabilize. In most parts of the country, a customer feels the greatest pinch of high fuel bills in the winter; by the time he or she decides to take action, figures out what to do, and then goes out and buys the visqueen or the wood stove, spring has arrived, the pressure is off, and yardwork seems to take precedence. Nine months later, however, the gas utility (proud of its success in holding the line on recent rate increases) will witness a stunning and wholly unexpected loss in sales volume.

## MARKETING PRACTICES AND PRINCIPLES

Faced with a loss in customers and per-customer sales, an interstate pipeline or a local gas distributor will, of course, suffer a decline in revenues. The outlook for future revenues and profits is gloomy unless the company takes decisive action.

There are three basic approaches to maintaining revenues in the face of a decline in sales. One approach is to simply charge higher per-unit prices for the sales volume that remains. Alternatively, a company can attempt to recover lost customers by shifting the cost burden disproportionately onto those who are least capable of abandoning the system. Both of these methods rely upon *cost allocation* rather than the third, and more direct approach of *cost reduction*. Gas producers are the most visible targets for cost cutting because wellhead-gas prices accounted for 62 percent of the end-use price of gas in 1982 and fully three-fourths of the cumulative increase in end-use prices since 1979.

### Changes in Per-Unit Gas Rates

Adjusting to increases at the wellhead by raising per-unit gas rates is consistent with the usual practices of gas utilities. Protected from the rigors of competition, franchised companies have traditionally operated on a "cost-plus" basis. When a gas utility invests in a new gas plant, or when employee salaries go up, it routinely files for a rate increase with its appropriate regulatory body.

Automatic add-ons to accommodate escalating wellhead prices were, in fact, institutionalized in the mid-1970s. Gas pipelines entering into "emergency gas purchases" (which circumvented the usual price restraints on wellhead-gas sales) swamped the Federal Power Commission with filings for rate revisions. Unable to keep up with the workload, the commission authorized the companies to incorporate *purchased-gas adjustment clauses (PGAs)* in their tariffs. PGAs made it possible to adjust the per-unit price automatically every six months to account for increases in gas-acquisition costs. The problem with this approach is the profoundly perverse effects it tends to have when it is used to deal with revenue shortfalls issuing in whole or in part from price-induced load losses, rather than from rising prices alone.

It is important to remember that the fundamental reason a loss in sales volume cuts into company earnings is the tendency for a dollar loss in sales to reduce system costs by less than a dollar. This is because only the *variable costs* are eliminated—such as the bookkeeping charges on a lost customer account, the fuel to compress the gas for shipment, and, most important, the acquisition price of the gas itself. The company cannot shed the *fixed costs* previously supported by the now-foregone sales. Specifically, the company still has to service the same amount of debt and recover its sunk equity investment in the same physical assets.

A rate increase seems to be in order, but this in itself is likely to discourage customer sales, especially if price escalation was the reason for load loss in the first place. Additional customers leave the system, and everybody else attempts to cut back on consumption, thus necessitating another rate increase almost as soon as the first one takes effect. (United Gas Pipeline Company, for example, was criticized by its downstream pipeline customers in 1983 for "pancaking" requests for rate increases; filing again and again for special permission to charge higher rates during the months between its semiannual PGA filings.) It is not difficult to see how this cycle might become self-perpetuating once it begins, plunging the utility into a *death spiral*.

The threat is not just theoretical. Urban transit after World War II provided a vivid example of a real-world death spiral. The flight of city-dwellers to the suburbs and the greatly expanded ownership of automobiles cut severely into transit demand. Bus and streetcar companies raised fares to make up for the revenue loss, which in turn alienated even more customers. The cumulative impact of repeated fare increases brought about the failure of almost every privately owned urban transit company in North America, and its replacement by a government-owned entity that requires a permanent subsidy.

A more immediate instance is the plight of electric utilities in the Pacific Northwest during the 1980s. They run the risk of entering a death spiral if they attempt to bill their customers for the cost of several nuclear power plants that were abandoned by the Washington Public Power Supply System (WPPSS, or "Whoops") after billions of dollars were invested in construction.

The gas industry is no less vulnerable. The Interstate Natural Gas Association of America (May 1983) reported that increased gas-transmission costs accounted for about 20 percent of end-user rate hikes during 1982, and fully half of that was attributable to load loss. The effect of load losses on gas-distribution companies was undoubtedly even greater.

### "Creative Accounting" and Utility Rate Structures

The second approach for maintaining company solvency in the face of declining sales and declining revenues is known by detractors as *creative accounting*. The burdens of escalating costs are simply shifted from one customer class to another or from present to future customers. (The latter is possible by skillful use of long-term financing that enables current cash costs to be incorporated in the company's rate base and amortized in the years to come.)

By early 1983, four interstate pipeline companies petitioned FERC for approval of revised tariffs that provided special "incentive rates" or "discounts" for industrial customers. Most of these companies later retracted their requests, after FERC approved the Columbia Gas Transmission Company tariff contingent upon utility stockholders rather than customers bearing the risks of undercollection. FERC's ruling made it clear that if the revenues from increased sales volume to price-sensitive customers failed to make up for the overall loss inherent in the per-unit rate decrease, federal regulators would be reluctant to approve offsetting rate increases to other customer classes.

The commission was concerned that by allowing companies simply to juggle their rates to maintain gas sales, gas pipelines would have a weaker incentive to seek price concessions from their suppliers at the wellhead. Northwest Central Pipleline Corporation was granted favorable terms for its incentive rate proposal, but the commission prefaced its ruling by observing that the beleaguered company was already doing just about everything it could to reduce its gas-purchase costs. Perhaps, too, the fact that Northwest Central officials were not personally responsible for their company's woes drew a measure of sympathy from the commissioners; Northwest Central was the result of Northwest Pipeline Company's takeover of the utility arm of the foundering Cities Service Gas Company.

State public utility commissions, on the other hand, had less justification for criticizing the gas-acquisition habits of their distributors. Most gas distributors received their entire supply from an inter- or intrastate pipeline. They had no voice in the producer contracts negotiated "upstream." More surprisingly, they had little direct control over the terms and conditions of their own purchase commitments to the pipeline company.

While a distributor did have the power to negotiate the terms and conditions of its *service agreement* with pipeline suppliers, those contracts typically referred to the pipeline's filed *tariff* for price terms (*rate schedules*) and delivery specifications. The pipeline, in turn, filed its tariff with FERC, and when FERC approved tariff revisions, these changes applied to all distributor service agreements—including those that predated the revision itself. Opportunities for distributors to make their concerns known at tariff hearings were, of course, available. But when the commission authorized pipelines to include PGA clauses in their tariffs, rate increases began to take effect automatically.

Most gas utilities and state regulatory commissions acquiesced in the approval of PGAs when gas was in short supply, but in 1983, after pipeline purchasing misjudgments had become painfully visible, the PGAs came under vigorous attack. Although the federal Constitution limits the ability of states to reach upstream into interstate commerce, they can make the situation politically uncomfortable for interstate pipelines and their federal regulators by striking PGAs from distributor tariffs within their jurisdiction. In the fall of 1982, Michigan voters did just that; however, because the ballot included two conflicting measures restricting the operation of PGAs and because both captured the necessary votes, implementation was delayed until the courts determined a course of action.

Aside from industrial discounts and incentive rates, both the FERC and state commissions are finding merit in revamping the fundamental structure upon which customer rates are designed. There is, after all, no economically "correct" way to apportion utility costs among the various classes of customers. More specifically, there is no inherent reason why all customers receiving gas from a particular pipeline or distributor ought to pay the same price. Indeed, there are a number of respectable theories that argue for nonuniform rates.

For example, because a transmission company must build a longer pipeline to carry gas from Texas to Connecticut than to Kentucky, pipeline tariffs universally reflect mileage disparities. Indeed, in 1957, the Federal Power Commission ruled that interstate pipeline tariffs must reflect cost differences between designated *zones* throughout a pipeline's market. Likewise, new customers within high-density city centers are much cheaper to connect than rural residents; distributors often bill their urban and rural customers accordingly.

Volume of use is another factor that calls for differential treatment of gas customers. The metering equipment and bookkeeping costs incurred per customer are relatively stable regardless of the volume of gas sold to each, and it can be argued that the large-volume customer should pay a somewhat lower per-unit rate. Sometimes the sheer volume of gas consumed by one or several industrial plants benefits nearby residential customers in a direct and undeniable fashion. Economies of scale fostered by large industrial loads result in per-unit system costs that are far lower than if the pipeline were scaled exclusively for residential needs. This is especially true for smaller communities distant from trunk lines, and promotional rates are an accepted practice for enticing added industrial consumption. The community gets its gas supply and more jobs, and the residential energy user benefits from both.

Utilities also find industrial customers valuable from the standpoint of *load management*. In most regions of the country, gas demand soars during the coldest months. If residential users comprised the bulk of gas sales, the utility would have to provide almost seven times more gas in the winter months than during the summer.

As a public utility, the company is directed to do all that it can to meet the needs of its customers—including their peak-day needs. A transmission or distribution company must, therefore, build a system large enough to accommodate anticipated winter demand. But by doing so, it is saddled with unused capacity during the off-season, which, in turn, means higher fixed costs per unit of gas sold.

This is where the industrial component of utility sales comes to the rescue. If a utility can count on a fairly steady year-round draw by industrial customers (for which weather-sensitive space heating is a minimal component of energy demand), the spread between average and peak-day use is reduced and more gas is carried overall. *Load factors* (the ratio between average and peak-day use) can be improved even more if industrial customers sign contracts that award utilities unilateral rights to cut off gas deliveries when peak demand of firm customers tests the limits of systemwide supplies.

*Interruptible contracts*, however, have little practical effect on load factors unless they are actually put to use. Until the 1970s very few interruptible customers had their gas turned off. Instead of maximizing load factors and scaling equipment accordingly, utilities chose to build gas pipelines and mains with an eye to future growth. Excess capacity in the present was an unavoidable cost for ensuring that a company could meet its customer-service obligations in the future. Nevertheless, interruptible contracts did improve utility load factors, albeit indirectly. To compensate for the risk of curtailment, interruptible contracts carried big per-unit discounts, and gas became an attractive fuel for industrial customers.

Historically, gas utilities have used *two-part tariffs* to apportion fixed and variable costs between firm and interruptible customers. Interruptible customers pay only a *commodity charge* (sometimes called *energy charge*) per unit of gas delivered. Firm customers pay a commodity charge commensurate with actual deliveries, but they also pay a *demand charge* in order to retain their right to draw gas "on demand" up to their contracted maximum. The commodity charge, in essence, covers the cost of the gas itself, while the demand charge buys a piece of the pipeline.

Apportioning costs between the commodity and demand components of the tariff is more a political art than an economic science. All of the variable costs that fluctuate with the volume of gas actually shipped (most important, the wellhead price) indisputably belong in the commodity charge. But how should fixed costs be split between the two parts of the tariff?

This question is more than academic; the greater the allocation of fixed costs to the demand charge (which interruptible customers do not pay), the greater is the inducement for industrial consumption of gas. Firm customers are certainly better off as long as the interruptibles pick up even a tiny portion of the fixed costs. On the other hand, there is no reason that industrials should be granted an exceptional bargain that would not have been possible absent the firm demand that justified pipeline construction in the first place.

Until 1952, the Federal Power Commission regulated pipeline rates in a manner that allocated all of the fixed costs to the demand component. In 1952, however, it sanctioned the *Atlantic Seaboard formula* (so named for the pipeline to which it first applied). The Atlantic Seaboard tariff structure split fixed costs equally (50/50) between the demand and commodity charges. That allocation standard remained in force until 1973, when federal regulators began to favor the *United formula*, which was even less advantageous to interruptible customers. The United formula allocated fully 75 percent of the fixed costs to the commodity share of the tariff paid by everybody. Throughout the gas shortages, the pipelines made repeated attempts (both administratively and in the courts) to resurrect the Atlantic Seaboard method, but they were consistently overruled.

The United formula was also adopted by state regulators enforcing similar ratemaking principles among distribution companies. By the late 1970s, price breaks for interruptible customers had no political constituency and, given the gas shortages, little economic merit. If natural gas was, indeed, fast approaching exhaustion, it made no sense to encourage low-priority uses by offering industrial consumers discounted rates. What is more, the architects of the incremental pricing provisions of the 1978 Natural Gas Policy Act were intent upon making industrial customers bear more than an equal burden of the higher cost of new supplies of natural gas. The two-part tariff, and the attendant discount for interruptible industrial customers, was, therefore, on death row.

Industrial gas customers suffered under several other governmental initiatives of the 1970s. Individual states outlawed *declining block rates* that had favored large-volume users for several decades. State legislatures, in fact, imposed just the opposite, reasoning that *inverted* rate structures encouraged conservation. California led the nation by adopting *lifeline rates* during the energy shortages, which reduced the per-unit charges for a base volume of monthly household deliveries that was deemed essential for human existence. But when gas prices began their upward climb in the early 1980s, maintenance of lifeline provisions forced industrial customers to bear a disproportionate share of escalating gas-purchase costs, and lifeline rates came under attack.

In 1982, with the specter of a devastating loss of industrial load, some companies advocated a return to Seaboard-style formulas or even more drastic structures that would widen the gap between per-unit costs to industrial and residential customers. Chicago's gas distributor, Peoples Gas Light and Coke Company, advocated a *fixed-variable* rate structure, which assigned all of the fixed costs to the demand component. Although FERC officials had been cool to pipeline proposals for emergency approval of special industrial discount programs, they voiced generic support for redesign of company rates along similar lines—provided that interested parties had the opportunity to voice their views via standard tariff proceedings. Finally in April 1983, a FERC administrative law judge broke precedent. He approved (subject to commission concurrence) a fixed-variable rate design for Natural Gas Pipeline Company of America.

Whenever and wherever the average cost of gas delivered to industrial customers—either directly by a pipeline or through a distributor—approaches or exceeds the price of residual oil, there will be a powerful drive for pricing gas according to its value to various customers. *Value-of-service* pricing (sometimes called "Ramsey pricing") can also be interpreted as a kind of price discrimination that is specifically disallowed by the Natural Gas Act of 1938 and antitrust laws. Advocates of rate revisions resembling value-of-service pricing must, therefore, base their arguments on other factors, such as the fixed versus variable cost considerations discussed previously. Free of the NGA standards, the Illinois Commerce Commission in mid-1983 approved an industrial incentive rate structure for Peoples Gas Light and Coke Company that was unabashedly labelled "value-of-service."

## Transactions at the Wellhead

The most direct methods for attacking the problems of swelling gas supplies and falling demand involve changes in supply arrangements—both volume and price. The first thing a pipeline drowning in surplus gas must do is eliminate the excess. How and where the company chooses to cut back bears a profound effect on the per-unit costs of the marketable gas.

On its face, the most rational curtailment policy would reduce takes from the costliest supplies first. In the early 1980s, however, interstate pipelines did not always do what seemed to be sensible. In fact, some pipelines found it necessary (and arguably in their consumers' best interest) to cut back disproportionately on their lowest-cost supplies. Why were market dynamics so confusing?

*Take-or-pay* obligations were largely at fault. As gas supplies tightened in the late 1960s and throughout the 1970s, producers were able to drive a hard bargain for sale of newly developed reserves. Especially since federal regulation prevented them from obtaining high prices, producers drew up sales contracts with other terms in their favor. Most notably, they were able to extract promises that the interstate pipeline purchasing the gas would always *take* a high percentage of the volume available (sometimes upwards of 90 percent). If the pipeline could not or would not meet the minimum threshold, it would nevertheless *pay* for the unclaimed volumes.

Not surprisingly, take-or-pay penalties were most frequent in recent (and costlier) supply agreements. The most debilitating obligations were those attached to the unregulated "deep" gas contracts signed by interstate pipeline companies during the four-year period that followed passage of the 1978 Natural Gas Policy Act, before the appearance of the gas glut. Coupled with price terms four or five times greater than the ceiling price for newly discovered shallow gas, deep-gas purchase contracts committed pipelines to take or pay for a high proportion of a well's delivery capacity.

On the other hand, many lower-priced contracts that were signed a decade or more earlier were still in force and did not impose such burdens. Thus, when faced with a greater commitment to wellhead purchases than their downstream markets could tolerate, interstate pipelines found that restricting their take of cheap old gas while maintaining their threshold takes from deep gas wells was in fact the strategy that balanced supplies with demand at the lowest net cost.

Take-or-pay burdens were exacerbated by other contractual provisions that, in essence, granted the producer complete control over gas *deliverability*. A pipeline not only had to take (or pay for) a big percentage of daily volumes that a specified reservoir would yield, but it was left up to the producer to determine the rate of flow. When gas prices increased, many producers decided to drill more development wells to increase deliverability on their properties. Then too, the Department of Interior had responded to the gas shortages of the 1970s by binding its Outer Continental Shelf lessees to accelerated development schedules. Once the investment was in place, OCS producers were sure to resist production cutbacks, even if the governmental mandate were relaxed. This financial incentive for high production rates was bolstered by the fact that OCS wells were outside the bounds of state conservation jurisdiction. Because the federal government had failed to relieve this regulatory gap, the Rule of Capture prevailed; gas belonged to whoever could bring it to the surface and find a buyer.

When the gas glut hit and pipelines were faced with a loss of industrial load (or, worse yet, the prospect of a death spiral), they reacted in a variety of ways. In the spring of 1982, Transco proclaimed that it would no longer buy deep gas at prices above $5.00 per mmbtu. Other companies followed suit, some declaring thresholds above and some below that of Transco. The timing of each pipeline's exit from the deep-gas purchasing frenzy also varied; some continued to sign contracts at prices upward of $8.00 per mmbtu as late as August of that year. But by the end of 1982, nobody was buying deep gas at prices above the ceiling of approximately $3.00 for still-regulated categories of gas; some pipelines were buying even regulated gas at below-ceiling prices; and others (most notably, El Paso Natural Gas Company) were not entering into any new contracts with anybody.

But most pipelines did not recognize the approaching marketability crunch soon enough. As a result, few companies could solve their problems simply by avoiding new high-cost contracts. Somehow they had to reduce volumes they had already committed to take. They "marketed-out" wherever contracts allowed; they cut back to threshold levels on take-or-pay wells; some drastically reduced their purchases from old-gas wells free of take commitments. Wherever possible, they tried to sell surplus gas *off system* to other pipelines that had not yet encountered the limits of customer demand. But by the time regulatory obstacles to off-system sales were overcome, the gas glut had infected the entire nation, and there were no regional shortfalls left to fill.

Even these actions failed to bring supply and demand into balance. Worse yet, many pipelines faced a dismal future if they could not retrieve some of the loss in load that had already occurred. In 1982, domestic producers found markets for only 17.7 tcf—a 10 percent decline from 1980 sales of 19.6 tcf. This load loss was far greater than what the recession alone could have induced, and for some pipelines it threatened their financial viability. It was time to take drastic action.

Columbia Gas Transmission Company led the way, boldly announcing in August 1982 that it would not take and could not pay for all the gas to which it was contractually committed. Lacking bona fide *market-out clauses*, Columbia simply claimed *force majeure* (which is a standard component of virtually all business contracts and a common-law principle to boot). Columbia argued that the "superior forces" of weather and the recession absolved it of its contractual responsibilities.

Other companies followed Columbia's lead, and enraged gas producers began to file suit. But the pipeline companies soon realized that they were in the driver's seat: *They* controlled the valves that released gas from producer wells, and *they* had the power to refrain from making payment. A producer could threaten to sell the reserves to somebody else, but pipelines knew that there were no other buyers in sight.

What is more, pipelines claiming force majeure were assured of at least a few years' grace period before a district court, and then an appeals court, and perhaps even the Supreme Court, would rule. Before that point, cash-squeezed producers might have agreed to drop charges (and drop prices) in order to reestablish gas production. Then, too, there was a good chance that FERC might do something to give the pipelines firmer legal ground for their ultimatums, and in 1983, Congress was seriously considering bills that would nullify take-or-pay commitments (at least up to a point).

Columbia's action was a good gamble. The business of abrogating take-or-pay commitments soon became an art, with industry leaders like InterNorth, Inc. finding ways to achieve the same result in a subtler, and perhaps more defensible manner. Of 1,300 producers advised by InterNorth that renegotiation was in order, only one had sought court intervention by mid-1983.

On the other hand, Tennessee Gas Transmission Company's abrupt and non-negotiable announcement of severe purchasing cutbacks created an uproar—and made every other pipeline's actions look reasonable by comparison. Virtually all companies directed their attorneys to scrutinize contracts. In the words of one pipeline company executive, they were "looking for anything, even a misplaced comma" that might give them some legal basis for avoiding their take-or-pay *deficiency payments*. Northwest Central Pipeline Corporation, out of desperation and under the shrewd guidance of John McMillian, launched some of the most ingenious attacks. The company finally broke through AMOCO Oil Company's refusal to relax stringent take-or-pay terms by finding a legally defensible reason to shut down some of

AMOCO's gas wells altogether: Northwest Central maintained that the gas did not meet the quality standards specified in the contract.

With domestic producers bearing reduced takes and increasingly acquiescing to reduced per-unit prices, gas imports (especially the significant volumes flowing from Canada) became a hotly debated issue on Capitol Hill. U.S. pipeline companies purchasing gas from Canadian suppliers found that their claims of force majeure went virtually unchallenged. And in exchange for a $400 million interest-free loan, Canadian producers agreed to reduce the minimum obligation for future shipments and to waive penalties on past take-or-pay infringements by the utilities that shipped Canadian gas through the Northern Border Pipeline. Meanwhile in the spring of 1983, the Canadian government dropped the border price from the long-established $4.94 per mmbtu to $4.40, with the prospect of further reductions to bring Canadian prices in line with market realities in the United States.

While all this activity and "market reordering" was taking place within the industry, FERC and Congress were trying to figure out what to do. Many of the proposals considered were either futile or dangerous; conditions were changing so quickly that the cumbersome bureaucratic and political processes simply could not do justice to a moving target, especially given a politically explosive climate that tended toward overreaction. By threatening abolition of PGAs and imposition of *contract carrier* obligations or, worse yet, *common carrier* status on interstate pipelines, FERC and Congress seemed determined to punish transmission companies for their past purchasing practices—practices intended by Congress in the Natural Gas Policy Act of 1978, recommended by the industry's consultants and commentators, approved by FERC, and (with few exceptions) endorsed by state regulatory commissions. In early 1983, gas pipelines had seemingly supplanted the hated oil majors in the role of primary economic scapegoat.

There was no rule, however, that FERC or Congress could devise which would add anything to the incentive that consumer resistance had already given pipelines to keep their gas-purchase costs down and their *plant factors* up. Gas companies would suffer far greater retribution for their purchasing errors at the hands of their own customers and the securities market than any penalty government regulators would dare to impose.

Under these circumstances it behooved pipeline companies to ensure that the volumetric and price terms agreed to in future contracts were sensible and that renegotiation of existing purchase agreements did not fall short of market realities. But what were these market realities? What was a market-clearing price for gas in the field today, and what would it be tomorrow?

## CALCULATING A MARKET CLEARING PRICE

One factor that caused much confusion in the deregulated environment following passage of the 1978 Natural Gas Policy Act, was the presence of an *old-gas subsidy cushion*. Because a pipeline traditionally *rolled in* the costs of its various gas purchases and charged downstream customers prices based upon the average, there was little question that a company could acquire some volume of overpriced gas and still be able to market its entire supply.

Perhaps the most scandalous statistic associated with the NGPA-induced market upheaval was the enormous spread between the highest- and lowest-cost field transactions. In 1982, some old gas was still flowing into the nation's pipelines at prices in the range of 26 cents per mmbtu, while deep-gas producers were pocketing almost fifty times as much for the same product. Transwestern Pipeline Company's purchase of Permian Basin gas for $10.10 per mmbtu marked the zenith of the deep-gas frenzy, but even the average price for deep gas was high. In 1982 deep-gas contracts averaged $7.33 per mmbtu.

Calculating a market-clearing price in 1978 or 1980 would have been difficult, even in hindsight, because of this old-gas cushion and the lack of any real-world examples by which to test hypotheses about industrial fuel-switching and the price elasticity of residential gas demand. By 1983, however, such was no longer the case. The subsidy cushion had been squashed out of sight, demand was on the decline, and there was little question that marginal purchases of gas would have to be acceptable to marginal customers. Quite simply, it made no sense for gas pipelines to sign any new contracts at prices higher than the btu-equivalent price of residual oil. And until the surplus subsided, they could not justify new purchases even at that price.

Although the manner in which interstate pipelines had traditionally gone about their business of buying and selling gas was constrained for the time being, the lingering surplus provided fertile ground for the birth of altogether new transactions. With cash-short producers hungry for markets, shut-in fertilizer companies prowling for a workable feedstock price, and dual-fuel industrial plants interested in bargains from anybody, it was not long before these potential buyers and sellers found one another.

Direct transactions between shut-in gas producers and industrial plants flourished in the intrastate markets of Texas and Louisiana, which, unlike interstate markets, had a long and respectable tradition of contract carriage by pipelines. (Statistics for 1980 indicate that one-third of all U.S. gas sales took place in intrastate markets, and of these, a third involved no intermediate purchase and sale by a pipeline.) More fundamentally, however, the direct transactions began to take on many of the characteristics of *spot markets* similar to those found in the oil business. Instead of long-term contracts at fixed prices with fixed escalators, some buy-and-sell commitments were of extremely limited duration, with prices renegotiable from month to month.

Even in the gas-importing states, direct sales and spot markets began to develop. With the exception of small volumes of local production that happened to be in close proximity to an industrial plant, negotiations with a major pipeline could not, however, be avoided. Unless a three- or four-party swap was arranged to move gas "on paper" (a standard practice within the crude-oil and refining business), the gas still had to be carried from buyer to seller.

In some cases, interstate pipelines willingly took on *contract carriage* business. They recognized that clinging to their traditional role as buyer and seller of gas was not as important as replenishing faltering loads. Transco, in fact, promoted contract-carrier arrangements by serving as a broker for prospective buyers and sellers. Transco's "industrial sales program" began in May 1983. Soon after, Panhandle Eastern implemented a similar program.

The onset of voluntary contract-carrier programs and the appearance of spot markets were the first structural manifestations of the changes unleashed by the Natural Gas Policy Act of 1978 and its de facto deregulation of domestic gas prices. (See Chapter 8 for a discussion of the industry's structural evolution.) In 1983, while it is impossible to foresee the course of future events in detail, it is safe to generalize on the direction of the changes that are in store. Fundamentally, the gas industry—producers and utilities both—will have to adjust to the opportunities and pitfalls of competition.

Columbia Gas Transmission Company learned the hard way that if it did not consent to move the gas that Sohio's fertilizer plant in Ohio had specially arranged to purchase, somebody else would. That somebody was Michigan Consolidated Gas Company. Changes are in store for distributors too. In October 1982, for example, the gas distributor serving Philadelphia was charging residential customers over $2.00 more per mmbtu than customers of another distributor in Pittsburgh had to pay. A situation like this is ripe for entrepreneurial innovation.

Because of the gas glut, producers have found not only that gas competes with residual oil for markets, but that gas competes with gas. Over the longer term, however, a persistent gas surplus is no more plausible than a persistent shortage. Prices are becoming increasingly free to move up *or down* to keep supplies and demand in balance. This is not to say that temporary imbalances will be unthinkable. As long as institutional and regulatory practices prevent the quick and clear translation of consumer inclinations back to producers in the field, and as long as new gas production is preceded by investment decisions several years old, the gas business may be doomed to cyclical swings between surplus and shortage.

*Market-demand prorationing* in the producing states can temper these swings in the business cycle, in unexpected weather patterns, and so forth. But with the exception of Texas (which has been rationing crude-oil production for several decades and worked the quirks out of its gas programs before the glut struck), the gas-exporting states will have a difficult time resolving the inevitable conflicts among producers and between producers and their pipeline customers. By mid-1983, in fact, the controversy over state prorationing was emerging as the most pressing frontier for regulatory action.

Should, for example, state oil and gas conservation boards (which in some states are the same entities that also regulate gas distributors) trim the deliverabilty excess by cutting back all producers equally on a *pro rata* basis? In addition to the usual production priorities granted casinghead gas (which, if prorationed could shut in oil production), should any other categories of gas wells receive special consideration? Should, perhaps, deep-gas producers in a cash squeeze be granted higher *allowables*? What about the prospects for instituting a "white market" by which producers could trade allowables among themselves, thus blending economic efficiency with pro rata equity? Finally, how much power should pipelines have to temper strict pro rata takes with their own intrrests in purchasing gas on a *least-cost* basis, and to what extent should

FERC assert jurisdiction over interstate commerce to ensure that cut-backs are handled in accordance with national policies?

Even if the gas industry as a whole miracuously works out all of its problems, ensures that price signals flow freely between the producer and the energy user, and banishes the entrenched attitudes of public utilities who have come to view government as a shield from the rules of the free-enterprise jungle, there is still one big and uncontrollable unknown that is bound to keep the industry on its toes (and watching *Platt's Oilgram*). Gas marketability is profoundly influenced by changes in world oil prices. Forecast of future market-clearing prices for gas are no more reliable than the forecasts of future oil prices upon which they are based.

Any serious student of the gas industry must, therefore, cultivate an understanding of the oil business. This chapter concludes with a rather detailed, but indispensable, discussion of crude-oil prices (adapted from articles originally published by the authors in 1983 issues of *The Public Interest* and *Medical Economics*).

## UNDERSTANDING WORLD OIL-PRICE TRENDS

Unlike natural gas, which is traded in distinct regions isolated by the high cost of waterborne commerce, there is a world price for crude oil. Quality variations and the expenses of port-to-port shipments do result in price differences, but these are usually slight relative to the average price. And unless a government chooses to insulate its consumers and industries from market realities by taxing or subsidizing imports, or restricting what its own energy producers may charge, crude-oil prices everywhere are about the same.

Beginning in the 1940s, the U.S. government did, in fact, implement programs that protected its energy producers from world-oil prices that it considered too low. After 1973 the concern shifted to consumers, and the government held the price of domestic production below the prevailing world price. These price ceilings then made it necessary to extend regulation even further. The government got into the business of allocating the cheaper domestic supplies rateably among all U.S. refiners through a complicated system of "entitlements."

Beginning in 1979, however, when President Carter ordered the gradual removal of petroleum price ceilings, the prices of crude oil produced in both the United States and various OPEC countries began to converge. By the end of 1982, there was a single worldwide price structure in which virtually all differentials could be explained by grade and quality or transportation cost differences.

After the two great oil upheavals of the 1970s, energy experts almost universally based their price forecasts on two assumptions: that oil prices would rise and continue to rise and that the Organization of Petroleum Exporting Countries (OPEC) would determine the rate of escalation—almost at whim. Beginning in 1981, however, *constant-dollar* oil prices entered their greatest decline in half a century. By mid-1983, the decline was still in progress; OPEC had apparently been able to stabilize *current-dollar* prices temporarily around the $29 official price for the Arabian light marker crude, but the new stability was a fragile one. Almost all forecasters had modified their short-term predictions of world oil prices to have them conform with the obvious, but few were willing to reject the wisdom of the past decade in formulating their long-term views.

Neither an end to the recession, nor OPEC attempts at production quotas, nor continued wars in the Middle East, however, will be able to shore up the crude-oil market permanently. The forces that led to the enormous price increases of the past decade can work just as effectively in reverse. It is, indeed, because oil prices climbed so rapidly and so high in the 1970s that they began in the 1980s to fall, and will likely fall further—perhaps as steeply and as far as they rose.

Although *real* (constant dollar) crude-oil prices may not rise to their 1980–81 levels again in this century, forecasting prices for any specific future year is a nearly hopeless task. History shows us no long-term oil-price trends but only a series of cycles of uneven duration and amplitude. In any event, the era of OPEC's opportunistic price gouging is over, and no other entity is in sight with the power to move oil prices in any consistent direction or to stabilize them at any given level.

### Oil Pricing before OPEC

To understand OPEC's helplessness in today's crude oil-market, it is useful to review how the market operated before OPEC came to power and how the Texas Railroad Commission (TRC) managed to exercise control for nearly forty years. The TRC's rule emerged in the mid-1930s from circumstances quite different from those that nurtured OPEC in the 1960s and 1970s. In the era between 1859 (the year Colonel Drake first discovered oil in Pennsylvania) and the Great Depression, crude-oil markets everywhere were dominated by events in the United States, where one black-gold rush after another unleashed an oversupply and sent prices plummeting.

Much of this market chaos resulted from the common-law *Rule of Capture*. The principle that nobody owned oil until it was brought to the surface generated frenzied competition among drillers to lift as much oil as they could from each newly discovered pool—before their neighbors did. The East Texas drilling rush ended in 1931 only when the governor sent the National Guard into the field to stop production.

The next year, a bitterly divided Texas legislature granted the TRC authority to limit output from individual wells in the interest of conservation and market order. Under *market-demand prorationing*, refiners told the TRC how much oil they wanted to buy each month, and the commission parceled out the "allowable" share of this demand to each well. This system assured every Texas producer a buyer for at least some of his oil, no matter how much excess producing capacity other producers held.

The TRC's ability to stabilize the market was bolstered by market-demand prorationing in several other states including Louisiana, the second-largest U.S. producer. Under state regulation, physical shortages and surpluses both became a thing of the past, conservation replaced physical waste, and the violent short-term fluctuations of crude-oil prices ended.

The TRC's effectiveness required a series of federal actions supporting its authority. In the 1935 Connally "hot-oil" law, Congress made it a federal crime to ship oil produced in violation of state conservation orders. After World War II, the executive branch acted to prevent uncontrolled imports of low-cost foreign crude oil from undermining the states' control of U.S. oil supplies. For a while, the handful of largely U.S.-based oil companies that controlled the oil reserves of the Middle East and the Caribbean cooperated successfully in limiting their foreign petroleum production to just about the amount demanded by their own foreign refineries.

Nevertheless, by 1948 the huge, low-cost oil reserves overseas so threatened Texas regulation that the Truman administration started assigning "voluntary" import quotas to the companies. In 1958, after independents like Hunt and Occidental developed enormous new reserves in Libya, President Eisenhower established a mandatory oil-import program (MOIP). The MOIP gave each U.S. refiner the right to import some lower-priced foreign oil, but it enabled the TRC and other state conservation authorities to continue setting the total volume of crude oil supplied to the domestic market.

This system of state regulation worked for nearly forty years, through World War II, the U.S. recessions of the 1950s, and several supply disruptions caused by Middle Eastern political upheavals. One reason for the TRC's success was that it did not exploit its market power opportunistically. The Middle Eastern conflicts in 1952, 1956, and 1967 offered Texas producers, the state of Texas (a major royalty owner), and the big international oil companies a chance for huge short-run profits. But each time, the TRC and the majors opted for long-term stability, forestalling the kind of consumer panic that generated the price run-ups of 1974 and 1979.

In hindsight, it was the TRC's very self-restraint that led to its downfall as world price maker. Because the TRC and its sister state commissions did not allow U.S. crude-oil prices to rise enough to offset the progressive depletion of low-cost domestic oil resources, they gradually stifled the incentive to find and develop new reserves. In the early 1970s, as a result, spare producing capacity in Texas and Louisiana dwindled to zero, and the state commissions were no longer able to accommodate even an orderly growth of demand, let alone to offset abrupt curtailments of Middle Eastern supplies as they had done in the 1950s and 1960s.

Once domestic production reached full capacity in 1972, the U.S. government had no choice—politically at least—other than to do away with import controls, leaving consumers exposed to whatever upheavals might occur in the oil-exporting countries. Meanwhile, nationalization of the major oil companies' overseas concessions, plus the growing influence of independents (including national oil companies like those of France, Italy, and Brazil), had stripped the majors of their ability to balance supply and demand outside of North America. A supply curtailment by the Arab oil producers, which would have hardly caused a ripple in oil prices ten years or even two years earlier, transformed world energy markets and, for a few years at least, handed control of those markets to OPEC.

The TRC determined prices by actively manipulating the aggregate supply of crude oil; as a state agency, it had the power to enforce its orders on the many thousands of Texas producers regardless of their conflicting individual interests and viewpoints. OPEC, on the other hand, has never had any authority over the diverse and sometimes warring sovereignties that make up its membership. Indeed, OPEC *per se* neither engineered nor enforced the decade's great price leaps; their immediate causes were the consumer panics that spread through the petroleum *spot market* after the 1973–74 Arab embargo and the 1978 Iranian revolution. In both cases, OPEC merely voted, after the panic had run its course, to establish the prevailing spot prices as the base prices for all crude-oil sales. It was a worldwide obsession with scarcity, rather than deliberate management of total world supplies, that fostered a mystique of OPEC invincibility and locked in the high prices the organization decreed in 1974 and 1979.

## A Decade of Panic Pricing

Oil producers and refiners usually try to plan their physical operations and to budget their purchase outlays and revenues well in advance. For this reason, the great bulk of the world's crude oil moves in "captive" channels from producing companies to their own refinery affiliates or on relatively long-term contracts between producers and refiners.

*Spot-transactions*—sales of a single tanker-load or less—usually account for only a small percentage of world supply, but they are an indispensable part of the total market because they allow any company or government to dispose of a temporary oversupply or fill a temporary shortfall. A general surplus or shortage equal to only, for instance, 3 percent of total world demand may thus show up as a surplus or shortage amounting to 50 or 100 percent of normal spot-market demand. As a result, spot prices tend to fluctuate daily and seasonally, and to range widely above and below *posted price* or contract price levels, which typically change slowly and infrequently.

Changes in crude-oil spot prices occasionally herald deep-seated market changes, but more often they are only exaggerated reflections of unexpected weather or business conditions, the buildup or drawdown of inventories, or political events. After some such contingency has caused spot prices to diverge sharply from contract prices, the spot market normally returns to a relatively narrow band of prices in the vicinity of previous contract-price levels. What was extraordinary about the OPEC-dominated markets of the 1970s is that they twice failed to respond in this normal way. After the panics of 1973–74 and 1979, spot prices did not fall back to precrisis levels; instead, contract prices rose—by OPEC decree—to the peak values to which the panic had carried spot prices. This feat was OPEC's great triumph—a triumph that, ironically, led to its downfall.

In 1974 the actual supply reduction was substantially less than the sum of (1) the oil then being consumed by electrical-generating and manufacturing plants that had the capacity to use other fuels, (2) the standby or underutilized oil-producing capacity of U.S. and uninvolved foreign producers, and (3) the inventory cushions that the industry ordinarily would have drawn down in order to prevent market turmoil.

In 1979, the panic and the runup of spot prices came first, and the reduction in oil shipments that validated the higher prices only afterward. Both surges in world prices began with a handful of large buyers who believed that the shortage was real or at least imminent and who were thus willing to pay almost anything. This crisis mentality had a powerfully perverse effect on the market. Instead of restraining demand, soaring spot prices gave the shortage credibility and helped propagate the panic to every class of consumer, so that demand actually increased.

Much of the apparent supply deficiency was caused by hoarding, the most memorable example of which was the fashion of topping off gasoline tanks in private automobiles on an almost-daily basis. Motorists, households, and businesses all sought to build up and maintain high inventories in case things "really got bad" later, while producers, refiners, and others expected to profit from holding products for resale at higher prices in the future. All of these anticipations were self-validating: Supplies did get tighter, and prices continued to rise.

### Roots of the OPEC Mystique

The spectacular spot-price rises of 1973–74 and 1979–80 could not have propagated themselves to the long-term sales arrangements if sellers and buyers had not both been captivated by the notion that resource scarcity made perpetually rising energy prices a fundamental law of nature. The ultimate root of OPEC's power was thus the doctrine that the world's fossil-fuel reserves were on the verge of exhaustion, which an unusually large number of parties embraced in the 1970s, albeit for different reasons.

Environmentalists hoped to slow the wasteful plunder of the earth's riches; oil companies were seeking to ward off price controls and attacks on their tax preferences; alternative-energy entrepreneurs sought business; politicians found in the energy crisis a moral equivalent of war; civil servants made it the rationale for massive expansion of their agencies and intervention into almost everything; and an army of academics, consultants, and journalists became rich and famous by studying, interpreting, or advocating national energy policies. Each group wanted to believe, or at least to persuade others, that "the wolf is really here."

This intellectual climate was an effective and durable weapon in OPEC's ideological arsenal. The organization had no enforcement machinery and did not even attempt to set production quotas for individual members until 1982, but the doctrine of tightening scarcity encouraged a sufficient number of OPEC members to reduce production when preservation of the price gains of 1974 and 1979 required it. The various countries made the required cuts individually, without coordination or urging by OPEC, because they believed that if they refrained from producing oil, its value would increase at a rate exceeding the plausible returns on cash investments of their oil taxes, royalties, or sales revenues.

## The Price of Overpricing

OPEC's hold over world energy markets in the 1970s, though mainly psychological, was no less real. However, the hold of the so-called cartel was far more fragile than the earlier market power of Texas, which stemmed from the TRC's direct control over production volumes. Today, few of the material requisites for further OPEC success remain. Its share of the world oil market fell from 55 percent in 1974 to barely one-fourth in mid-1983; and Saudi Arabia's share was less than half the share Texas held as late as the mid-1960s.

Some recognition of these shifting realities began to strike the Saudi leadership after two deliberate production cuts in 1979 had locked in the huge price increases voted by OPEC the previous year. But Saudi Arabia was too late in discovering the responsibilities that accompany the market power it possessed. Contrary to the near-perfect consensus of industry, government, and the academic-consulting community during the 1970s, crude-oil demand does respond—slowly but massively—to price changes. In the long run, higher prices have a profound effect on oil supply as well.

Non-OPEC output has grown rapidly and will continue to grow. Production from the North Sea, Alaska, and Mexico, for example, increased by 4 million barrels per day between 1977 and early 1982, and Mexico's exports—driven now by economic necessity—could conceivably increase by several millions of barrels per day before 1990. Most clearly and most importantly, however, high oil prices are shrinking oil demand, both by reducing total energy consumption and by making coal, natural gas, nuclear power, and other energy sources more attractive. An absolute decline in U.S. oil consumption was first visible in 1979; the rest of the industrialized world followed a year later.

In retrospect, it is remarkable how many experts believed that oil demand is insensitive to price changes. Except in a couple of OPEC countries, no new base-load generating plants fired by oil, or large-scale oil-fired boilers of any sort, have been built since the mid-1970s. And until the oil-price decline of 1981 was accepted as more than just a passing aberration, industry had been relentlessly converting existing oil-burning equipment to coal, natural gas, and other energy sources.

Because changes in the world's fuel-use patterns are generally embodied in long-lived capital-intensive investments such as buildings, transportation equipment, and industrial machinery, the lag between the big 1974 price rises and the onset of the absolute decline in oil consumption only reflected the time it required to replace those assets. This delay in adjusting the world's capital stock to changed energy-supply conditions also suggests that the high OPEC prices of 1974–83 will influence consumption patterns for many more years, even if prices continue to fall as rapidly and as far as they rose in the 1970s.

It should be fairly obvious now that predictions of $100 per-barrel oil are ludicrous. A world that is already fleeing from oil at $29 per barrel would hardly have any use for it at two or three times that price. When the price of oil passed $20 per barrel (in 1983 dollars), it became more expensive to burn as an industrial bulk fuel than U.S.-produced coal at any tidewater location in the world. And at prices in the $50 range, oil would begin to price itself out of even its most "captive" markets. Given a few years for market and infrastructure development, liquid petroleum products would become marginal even as transportation fuels, which increasingly would be replaced by some combination of compressed and liquefied hydrocarbon gases and alcohols.

### The New Buyer's Market

There is little prospect that OPEC can function effectively in a chronic buyer's market, especially in the face of the organization's current internal dissensions. The downward pressure on prices resulting from shrinking demand for OPEC oil is exacerbated by the shaky financial condition of the exporting countries, a drastic turnaround from the situation of the mid-1970s.

Since 1973, the OPEC nations' spending for imports has risen at an average annual rate of 30 percent because of ambitious industrialization plans and, in several cases, extravagant purchases of military hardware. In 1982, the combination of declining oil demand and rapidly rising expenditures had already resulted in trade deficits for all but three members of the cartel. Unless oil production or prices increase sharply, every member, including Saudi Arabia, will suffer deficits in the mid-1980s. These deficits, exacerbated by the continuing Iran-Iraqi war, inexorably drive the most hard-pressed countries (in search of revenues to pay for today's imports) to produce as much oil as they can sell, notwithstanding the price consequences.

Another big disturbance in world oil markets could push prices either up or down. It is still conceivable, if only barely, that a sharp economic upturn and an exceptionally cold winter could combine with the right kind of Middle Eastern political crisis and send prices soaring for a third time, to levels significantly above those reached in 1980–81.

The probabilities, however, weigh heavily on the other side. There is a huge overhang of excess production capacity in the oil-exporting countries. Several of them are in extreme fiscal distress; Mexico in particular has both the ability and a desperate need to increase oil exports. Meanwhile, the price-induced flight from oil is still gathering momentum (notwithstanding an unusual aberration in U.S. prices for natural gas that, beginning in 1982, prompted a shift to residual oil among boiler-fuel users). That momentum will not be spent for years, no matter what happens to oil prices today.

All of these forces together (not to mention the sluggish world economy) exert pressures that favor further price reductions. The serious questions are whether the remainder of the current cyclical price decline will be orderly or chaotic, and how far prices will ultimately drop before stabilizing and turning upward again.

## Is There a Long-Term Price Equilibrium?

Worldwide scarcity and rising real *resource cost* (capital, material, and labor costs—also called *economic cost*) had little or no direct responsibility for the energy-price upheavals of the 1970s. The earth's known resources still include plenty of crude oil that could be developed and produced at costs well below 1973 real prices. Considering these resources alone, there is enough low-cost oil left to satisfy the world's current rate of consumption for several decades. And for the reasons discussed in Chapter 6, *known* volumes do not delimit the bounds of resource availability.

The world's supply and demand for petroleum are more predictable than the prices that will prevail at any particular date over the next few years. Over a term of many years, the average price of oil can probably be forecast with reasonable confidence, but it is questionable how much value such a forecast would have in the absence of some institutional mechanism capable of stabilizing prices in the vicinity of such a long-term *equilibrium price*. OPEC can not be that mechanism, and no other effective producer cartel is in sight. If world oil prices are to achieve reasonable stability over the rest of this century, the stabilizing influence must be a greater flexibility in oil demand, rather than the kind of supply manipulation at which the Texas Railroad Commission was successful.

The only sustainable oil-price level is thus one wherein oil shares the world's boiler-fuel and bulk-fuels market with coal and gas. Oil prices cannot diverge substantially from the cost of mining, transporting, and burning coal in an environmentally acceptable fashion. Within this spectrum, the ability of large industrial plants to meet price changes by switching to a cheaper fuel—greatly enhanced on a global scale by the last decade's unforeseen price upheavals—could conceivably prevent shocks to the market (from either the supply or the demand side) from triggering the kind of panic behavior that might otherwise send prices soaring to $50 per barrel or sinking to $5.

The band of plausibly sustainable prices is nevertheless a broad one, ranging from about $12 to about $20 per barrel, or $2.00 to $3.50 per mmbtu, in 1983 prices. History offers some empirical support for the viability of a long-term average oil price in the $12 to $20 range. Over the past 110 years, the average price of crude-oil refined in the United States has been about $14 per barrel, and, despite an average constant-dollar price fluctuation of more than 20 percent per year, there has been no statistically significant trend in price, either upward or downward.

What does this outlook mean for natural gas? Two key inferences can be drawn from this analysis of oil-price behavior. First, if gas producers, pipelines and distributors are going to maintain or increase their share of the U.S. industrial market for energy, gas prices will have to retain the short-term flexibility to track oil prices—which, unfortunately, are notoriously unstable. Second, because oil prices in the long-term will be held in check by the prices that coal producers can offer industrial customers, long-lived investments in facilities to produce or deliver gas at a burner-tip cost that exceeds the cost of coal (with due allowance for the additional capital outlays required to deal with the environmental problems coal presents) are probably unwise. Stating this rule is easier than giving it a precise dollar value, but from a mid-1983 perspective, the long-term equilibrium value of gas delivered into the existing North American pipeline system is probably on the order of $3.00 per mcf (or mmbtu)—give or take a dollar.

Will the gas industry be able to adjust to changing oil prices over the short term, and will the cumulative investment decisions of its individual sectors (producers, pipelines, and distributors) ensure a stable or growing long-term market share? The answers depend powerfully upon the structure of the industry itself. Chapter 8 explores the institutions that have determined the structure of the gas industry to date and those that will likely influence it in the future.

# 8 STRUCTURAL EVOLUTION OF THE GAS INDUSTRY

## AN ARRAY OF DISPARATE COMPANIES

In contrast to the oil industry, production, transmission, and distribution of natural gas is characterized by a lack of vertical integration. With few exceptions, companies that are dominant forces in one sector occupy a relatively minor (or even nonexistent) position in one or both of the other sectors. Table 8–1 (with supporting data shown in Tables 8–2 through 8–4) illustrates this point. It identifies the top twenty companies in each sector of the gas industry and notes their affiliations.

Only one company, Columbia Gas Transmission Company, is a major force in all three sectors. Even so, Columbia's strength is primarily in gas transmission and distribution. An affiliate in Ohio is one of the top twenty retail distributors nationwide. Columbia also owns distribution companies in six other states, making its companywide distribution business perhaps the largest in the nation. As a transmission enterprise, its volume of gas sales and systemwide pipeline mileage is second only to El Paso Natural Gas Company.

Columbia is by no means the Exxon of the gas industry. The company name is virtually unknown in the West, since Columbia's transmission and distribution operations are confined to areas east of the

**Table 8-1.** Affiliations among the Top Twenty Gas Producers, Distributors, and Transmission Companies.

| Parent | Production | Interstate Transmission | Retail Distribution |
|---|---|---|---|
| **Fully integrated:** | | | |
| Columbia Systems | Columbia Gas | Columbia Gas Transmission | Columbia Gas of Ohio |
| **Major companies in production and transmission:** | | | |
| Tenneco | Tenneco | Tennessee Gas Transmission | None |
| | | Midwestern Gas Transmission | |
| El Paso | El Paso | El Paso Natural Gas | None |
| Panhandle Eastern | Panhandle Eastern | Panhandle Eastern Pipe Line | None |
| | | Trunkline Gas Company | |
| **Major companies in transmission and distribution:** | | | |
| InterNorth Inc. (formerly Northern Natural) | | Northern Natural Gas | People's Gas (a division) |
| Pacific Gas & Electric Co. | | Pacific Gas Transmission | Pacific Gas & Electric Co. |
| Consolidated Natural Gas | | Consolidated Natural Gas | East Ohio Gas |
| Peoples Energy Corp.[a] | | Natural Gas Pipeline Company of America | People's GasLight & Coke |
| American Natural Resources[a] | | Michigan Wisconsin Pipe Line | Michigan Consolidated |

**Single Sector Companies:**

| Production | Interstate Transmission | Retail Distribution |
|---|---|---|
| Exxon | United Gas Pipeline | Southern California Gas |
| Texaco | Texas Eastern Transmission[b] | Northern Illinois Gas |
| Standard of Indiana | Transwestern Pipeline | Lone Star Gas |
| Mobil | Algonquin Gas Transmission | Oklahoma Natural Gas |
| Gulf | Transcontinental Gas Pipe Line | Consumers Power |
| Shell | Southern Natural Gas | Arkansas-Louisiana |
| Atlantic Richfield | Texas Gas Transmission | Northern Indiana Public |
| Union | Northwest Pipeline | Service Co. |
| Standard of California | Colorado Interstate | Gas Service Co. (Missouri) |
| Standard of Ohio | | National Fuel Gas |
| Sun | | Atlanta Gas Light |
| Phillips | | Entex |
| Getty | | Southern Union Gas |
| Superior | | Pioneer Natural Gas |
| Conoco | | |
| Pennzoil | | |

Source: American Gas Association, *Gas Energy Review* (May 1981).

[a]These two companies spun off all their remaining distribution subsidiaries in 1981, including those listed on this chart.

[b]Although Texas Eastern is not a major gas producer, more than half of its revenues are from gas production.

# Table 8-2 The Top Twenty Producers of Natural Gas, 1979.

## U.S. Natural Gas Production, 1979 (Millions of Cubic Feet)

| Company | Production |
| --- | --- |
| Exxon | 1,462,555 |
| Texaco | 1,227,495 |
| Standard of Indiana | 903,010 |
| Mobil | 742,045 |
| Gulf | 692,916 |
| Shell | 634,370 |
| Atlantic Richfield | 486,910 |
| Union | 438,438 |
| Standard of California | 433,283 |
| Tenneco | 411,579 |
| Sun | 397,485 |
| Phillips | 391,280 |
| Getty | 318,547 |
| Cities Service (now Occidental Petroleum Co.) | 316,127 |
| Superior | 313,127 |
| Conoco | 304,410 |
| El Paso | 291,500 |
| Pennzoil | 214,288 |
| Columbia Gas | 166,468 |
| Panhandle Eastern | 158,025 |

## Proven U.S. Natural Gas Reserves, 1979 (Billions of Cubic Feet)

| Company | Reserves |
| --- | --- |
| Exxon | 17,200.0 |
| Atlantic Richfield | 13,354.0 |
| Texaco | 9,583.0 |
| Standard of Indiana | 8,654.0 |
| Shell | 6,703.0 |
| Standard of Ohio | 6,672.0 |
| Mobil | 6,498.0 |
| El Paso | 6,174.0 |
| Union | 6,038.0 |
| Standard of California | 5,300.0 |
| Gulf | 5,052.0 |
| Phillips | 3,656.0 |
| Sun | 3,200.0 |
| Cities Service (now Occidental Petroleum Co.) | 3,197.0 |
| Tenneco | 3,166.5 |
| Conoco | 2,681.0 |
| Getty | 2,546.0 |
| Marathon | 2,176.6 |
| Superior | 1,891.0 |
| Panhandle Eastern | 1,480.9 |

| Percentage of total production represented by | 1975 | 1979 | 1975-1979 |
| --- | --- | --- | --- |
| Top 4 | 24.2 | 21.3 | -2.9 |
| Top 8 | 36.8 | 32.3 | -4.5 |
| Top 15 | 52.2 | 45.0 | -7.2 |
| Top 20 | 58.1 | 50.6 | -7.5 |

| Percentage of total reserves represented by | 1975 | 1979 | 1975-1979 |
| --- | --- | --- | --- |
| Top 4 | 26.9 | 25.0 | -1.9 |
| Top 8 | 39.7 | 38.4 | -1.3 |
| Top 15 | 56.9 | 53.6 | -3.3 |
| Top 20 | 64.0 | 59.1 | -4.9 |

Source: American Gas Association, *Gas Energy Review* (May 1981). Adapted from American Petroleum Institute, Market Shares and Individual Company Data for U.S. Energy Markets (October 1980).

**Table 8-3.** The Top Twenty Interstate Gas-Transmission Companies, 1979.

| Company[a,b] | 1979 Sales (Trillion Btus) | 1979 End-Use Sales (Trillion Btus) | 1979 Total Pipeline System[c] (Miles) |
|---|---|---|---|
| El Paso Natural Gas Co. | 1,233.1 | 94.7 | 20,616 |
| Columbia Gas Transmission Corp. | 1,219.2 | — | 19,417 |
| Tennessee Gas Transmission Co. | 1,120.8 | 1.0 | 13,481 |
| Natural Gas Pipeline Co. | 1,014.3 | 9.0 | 11,855 |
| United Gas Pipe Line Co. | 962.9 | 215.3 | 9,119 |
| Texas Eastern Transmission Corp. | 910.7 | — | 9,242 |
| Transcontinental Gas Pipe Line Corp. | 738.0 | 2.6 | 9,290 |
| Michigan Wisconsin Pipe Line Co. | 729.9 | — | 10,205 |
| Panhandle Eastern Pipe Line Co. | 678.2 | 38.2 | 11,446 |
| Southern Natural Gas Co. | 639.6 | 47.8 | 7,154 |
| Texas Gas Transmission Co. | 623.8 | 5.4 | 5,857 |
| Houston Natural Gas Corp. | 514.9 | 271.9 | 4,005 |
| Trunkline Gas Co. | 504.1 | 0.2 | 4,331 |
| United Texas Transmission Co. | 473.2 | 380.0 | 1,393 |
| Northwest Pipeline Corp. | 447.4 | 4.9 | 5,491 |
| Cities Service Gas Co. (now, Northwest Central Gas Co.) | 400.7 | 106.1 | 9,167 |
| Pacific Gas Transmission Co. | 384.4 | — | 664 |
| Midwestern Gas Transmission Co. | 342.6 | — | 915 |
| Transwestern Pipeline Co. | 293.9 | 0.9 | 3,562 |
| Pacific Lighting Service Co. | 245.9 | — | 1,015 |

Source: American Gas Association, *Gas Energy Review* (May 1981).

[a]Data for the Colorado Interstate Gas Co., also considered to be a large transmission company, was not available.

[b]Transmission companies are defined by A.G.A. as being those companies that sell over 95 percent of their supplies to either other gas utilities or large industrial customers. Transmission companies must also classify over 95 percent of their pipeline mains as either gathering, storage, and/or transmission mains.

[c]Includes field and gathering, underground storage, and transmission mains.

**Table 8-4.**  The Top Twenty Gas-Distribution Companies, 1979.

| Company | State(s) Served | 1979 Sales Trillion of Btu | | | 1979 Customers (000) | |
|---|---|---|---|---|---|---|
| | | Total | End-Use | Residential | End-Use | Residential |
| Southern California Gas Co. | CA | 898.0 | 794.8 | 323.6 | 3,674.3 | 3,475.6 |
| Pacific Gas and Electric Co. | CA,UT,WA | 852.9 | 814.0 | 242.2 | 2,763.4 | 2,591.5 |
| InterNorth, Inc. | CO,IL,IA,KS,MI,MN,MO,MT NB,NM,OK,SD,TX,WI,WY | 748.9 | 206.1 | 37.4 | 276.0 | 239.9 |
| Consolidated Gas Supply Corp. | LA,NY,OH,PA,WV | 671.5 | 45.5 | 15.1 | 109.4 | 101.2 |
| Northern Illinois Gas Co. | IL | 537.5 | 537.5 | 221.8 | 1,407.8 | 1,283.8 |
| Lone Star Gas Co. | OK,TX | 507.9 | 466.7 | 95.7 | 1,146.8 | 1,042.2 |
| Michigan Consolidated Gas Co. | MI | 457.5 | 452.9 | 183.3 | 1,034.8 | 963.4 |
| The East Ohio Gas Co. | OH | 383.0 | 383.0 | 174.4 | 963.3 | 908.8 |
| Columbia Gas of Ohio, Inc. | OH | 359.0 | 358.6 | 151.6 | 1,030.8 | 954.0 |
| Oklahoma Natural Gas Co. | KS,OK,TX,WY | 335.9 | 277.7 | 61.4 | 569.4 | 516.5 |
| Peoples Gas Light and Coke Co. | IL | 328.2 | 326.9 | 175.2 | 889.4 | 743.4 |
| Consumers Power Co. | MI | 327.5 | 326.8 | 167.2 | 1,069.8 | 998.2 |
| Arkansas-Louisiana Gas Co. | AK,LA,OK,TX | 306.5 | 271.9 | 67.7 | 684.3 | 615.9 |
| Northern Indiana Public Service Co. | IN | 263.0 | 263.0 | 75.4 | 504.5 | 466.3 |
| Gas Service Co. (Missouri) | KS,MO,NB,OK | 237.9 | 237.9 | 120.1 | 823.2 | 775.2 |
| National Fuel Gas Co. | NY,OH,PA | 233.8 | 212.2 | 106.9 | 663.1 | 626.1 |
| Atlanta Gas Light Co. | FL,GA | 224.4 | 224.4 | 71.6 | 768.6 | 706.4 |
| Entex, Inc. | LA,MS,TX | 199.3 | 199.3 | 87.8 | 1,170.6 | 1,074.0 |
| Southern Union Gas Co. | AZ,AK,CO,NM,OK,TX | 196.5 | 166.1 | 50.1 | 584.8 | 530.0 |
| Pioneer Natural Gas Co. | TX | 195.9 | 153.8 | 27.6 | 264.7 | 228.3 |

Source: American Gas Association, *Gas Energy Review* (May 1981).

Mississippi. Even in the East, there are substantial markets never penetrated by the Columbia system. (See Map 8-1.)

Bigness is not everything, however. In 1982, for example, Columbia was singled out by the media as a vivid example of strategic misjudgments in its gas-acquisition practices. Dangerously overcommitted with take-or-pay and high-cost gas purchase arrangements, and amortizing a now-mothballed LNG facility, Columbia's rates soared in 1982. Higher rates prompted fuel switching and exacerbated the recession-fed epidemic of factory closures, depriving the company of a vital chunk of industrial load. Desperate, Columbia was the first of the interstate gas pipelines to walk away from its purchase commitments, and angry suppliers responded with a barrage of lawsuits.

Tenneco's interstate gas-transmission affiliates (Tennessee Gas Transmission and Midwestern Gas Transmission) own about twice as much large-diameter (20-inch and over) pipeline as the next biggest transmission company. Table 8-5 lists these affiliations, along with those of other pipeline companies. No other company engaged in gas transmission produces more gas, and Tenneco is a leader with respect to ownership of reserves as well. Even so, Tenneco does not reign over the gas industry. It has no retail distribution operations whatsoever, and, like Columbia, it has not penetrated the western half of the nation. Even in the East, there are substantial gas markets not yet tapped by the Tenneco system. (See Map 8-2.) The only transmission company, in fact, with coast-to-coast operations is Texas Eastern Transmission Corporation, with its affiliate, Transwestern Pipeline Company. (See Map 8-3.)

Table 8-1 illustrates two additional characteristics of the gas industry. First, the major gas producers are really the major oil companies—and the major oil companies have no transmission or distribution affiliates. Second, of the 1,600-plus distribution companies operating nationwide, even the largest are virtual unknowns outside of the (usually) single state served by each. Table 8-6 and Figure 8-1 display the extent to which gas-transmission companies nationwide have gotten into the production business. The statistics clearly indicate the lack of vertical integration between these two sectors.

The sectoral isolation of U.S. companies now engaged in gas production, transmission, or distribution is striking. History shows us, however, that this was not always the case. A number of factors have contributed to the disintegration of the gas industry. This chapter will survey the most important ones.

Map 8-1.   Columbia Gas Transmission Company

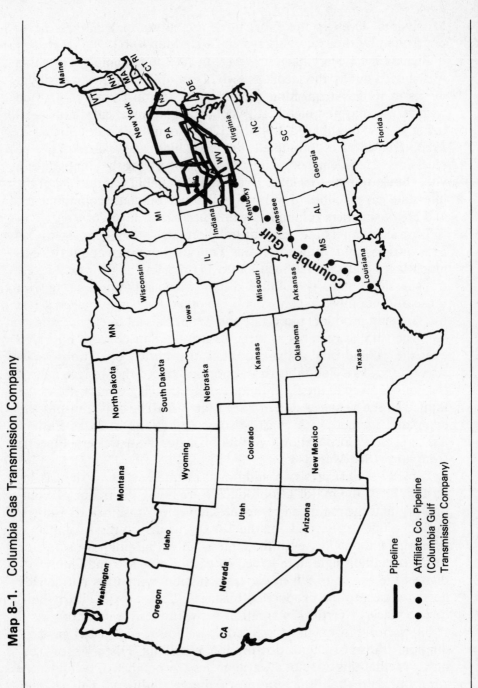

Pipeline

Affiliate Co. Pipeline
(Columbia Gulf
Transmission Company)

**Table 8–5.**   Corporate Affiliations of the Major Interstate Pipeline Companies.

| Affiliated Companies | Parent |
|---|---|
| Tennessee Gas Transmission Co. (100 percent) Midwestern Gas Transmission Co. (100 percent) | Tenneco, Inc. |
| Michigan Wisconsin Pipe Line Co. (100 percent) Great Lakes Gas Transmission Co. (50 percent) | American Natural Resources Co. |
| Panhandle Eastern Pipe Line Co. (100 percent) Trunkline Gas Co. (100 percent) | Panhandle Eastern |
| Texas Eastern Transmission Corp. (100 percent) Transwestern Pipeline Co. (100 percent) Algonquin Gas Transmission Co. (28 percent) | Texas Eastern Corp. |
| Columbia Gas Transmission Corp. (100 percent) Columbia Gulf Transmission Corp. (100 percent) | The Columbia Gas Systems, Inc. |
| Northwest Pipeline Corp. (100 percent) Northwest Central Gas Co. (100 percent) | Northwest Energy Co. |
| Pacific Gas & Electric Co. (100 percent) Pacific Gas Transmission Co. (affiliate) | Pacific Gas & Electric Co. |

**Map 8-2.    Tennessee Gas Transmission Company**

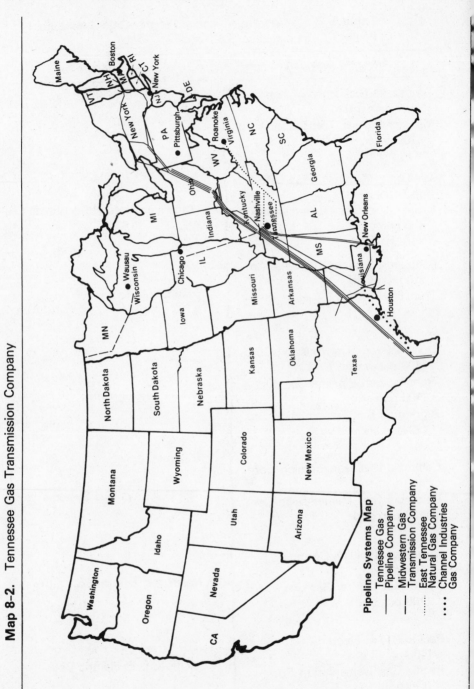

**Pipeline Systems Map**
— Tennessee Gas Pipeline Company
—| Midwestern Gas Transmission Company
⋯⋯ East Tennessee Natural Gas Company
•••• Channel Industries Gas Company

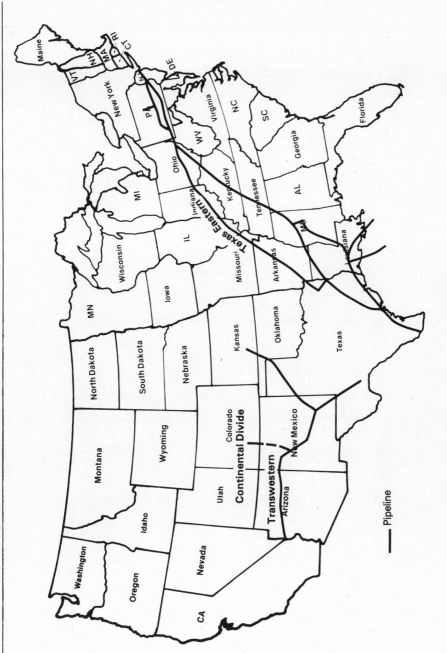

**Map 8-3.    Texas Eastern Transmission Corporation.**

**Table 8-6.** Gas Supplies of Major Interstate Transmission Companies, 1980.

| Pipeline Purchaser Name | Annual Supply Volume Billion cf | Owned Production | Producer Contracts | Percentage of Total Supply[a] — Purchases From Interstate Pipelines Seller Number[b] | | | | | | | | | | | | | Other Pipelines | Import |
|---|---|---|---|---|---|---|---|---|---|---|---|---|---|---|---|---|---|---|
| | | | | 2 | 7 | 9 | 10 | 12 | 13 | 14 | 15 | 16 | 17 | 18 | 19 | 20 | | |
| 1 Cities Service | 412 | 0 | 76 | 0 | 0 | 0 | 0 | 0 | 0 | 0 | 0 | 0 | 0 | 14 | 0 | 0 | 10 | 0 |
| 2 Colorado Interstate | 409 | 21 | 69 | NA | 0 | 0 | 0 | 0 | 0 | 0 | 0 | 0 | 0 | 0 | 9 | 0 | c | c |
| 3 Columbia | 1,121 | 4 | 46 | 0 | 0 | c | 0 | 7 | 18 | 0 | 12 | 8 | 2 | 0 | 0 | 0 | 1 | 2 |
| 4 Consolidated | 638 | 5 | 19 | 0 | 0 | 0 | 0 | 0 | 0 | 25 | 33 | 14 | 0 | 0 | 0 | 0 | 0 | 3 |
| 5 El Paso | 1,291 | 21 | 67 | 0 | 0 | 0 | 0 | 0 | 0 | 0 | 0 | 0 | 0 | 0 | 0 | 0 | 11 | 1 |
| 6 Florida | 186 | 0 | 80 | 0 | 0 | 0 | 0 | 0 | 18 | 0 | 0 | 0 | 0 | 0 | 0 | 0 | 0 | 2 |
| 7 Kansas-Nebraska | 121 | 32 | 57 | 5 | NA | 0 | 4 | 0 | 0 | 0 | 0 | 0 | 0 | 0 | 0 | 0 | 2 | 0 |
| 8 Michigan-Wisconsin | 691 | 1 | 80 | 0 | 0 | 0 | 0 | 0 | 0 | 0 | 0 | 2 | 0 | 0 | 0 | 0 | 12 | 2 |
| 9 Natural | 1,016 | 3 | 83 | 7 | 0 | NA | 0 | 0 | 0 | 0 | 0 | 0 | 0 | 0 | 0 | 1 | 6 | 0 |
| 10 Northern | 800 | 2 | 95 | 1 | 0 | 0 | NA | 0 | 0 | 0 | 0 | 0 | 0 | 0 | 0 | 0 | 1 | 1 |
| 11 Northwest | 353 | 11 | 34 | 0 | 0 | 0 | 0 | 0 | 0 | 0 | 0 | 0 | 0 | 0 | 0 | 0 | 0 | 54 |
| 12 Panhandle Eastern | 732 | 0 | 67 | 0 | 1 | 0 | 1 | NA | 0 | 0 | 0 | 0 | 0 | 0 | 31 | 0 | 0 | 0 |
| 13 Southern | 614 | 2 | 64 | 0 | 0 | 0 | 0 | 0 | NA | 0 | 0 | 0 | 0 | 0 | 0 | 8 | 25 | 1 |
| 14 Tennessee | 1,147 | 1 | 92 | 1 | 0 | c | 0 | 0 | 2 | NA | 0 | 0 | 0 | 0 | 0 | 0 | 3 | 4 |
| 15 Texas Eastern | 981 | c | 60 | 0 | 0 | 0 | 0 | 0 | 0 | 0 | NA | 11 | 0 | 0 | 0 | 25 | c | 2 |
| 16 Texas Gas | 420 | c | 99 | 0 | 0 | 0 | 0 | 0 | 0 | 0 | 0 | NA | 0 | 0 | 0 | 0 | 0 | 0 |
| 17 Transcontinental | 919 | 0 | 96 | 0 | 0 | 0 | 0 | 0 | 0 | 0 | 0 | 0 | NA | 0 | 0 | 0 | 2 | 2 |
| 18 Transwestern | 335 | 0 | 97 | 0 | 0 | 0 | 0 | 0 | 0 | 0 | 0 | 0 | 0 | NA | 0 | 0 | 0 | 3 |
| 19 Trunkline | 493 | 0 | 100 | 0 | 0 | 0 | 0 | 0 | 0 | 0 | 0 | 0 | 0 | 0 | NA | 0 | 0 | 0 |
| 20 United | 1,038 | 0 | 87 | 0 | 0 | 0 | 0 | 0 | 0 | 0 | 0 | 0 | 0 | 0 | NA | NA | 13 | 0 |

Source: Energy Information Administration, U.S. Department of Energy, *Gas Supplies of Interstate Natural Gas Pipeline Companies, 1980,* DOE/EIA-0173(80) (Washington, D.C., December 1981).

[a] Percentages may not sum to 100 due to independent rounding.
[b] Seller number cross-references to purchaser number given in left margin.
[c] Less than 0.5 percent.

Note: NA = not applicable.

**Figure 8-1.**   Gas Production by Pipeline Affiliates, 1982

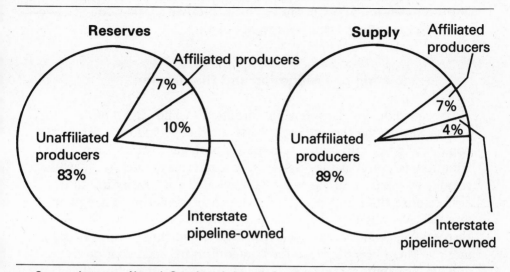

Source: Interstate Natural Gas Association of America. "Contract Issues Survey." Washington, D.C.: INGAA, May 1983.

## THE INFANT GAS INDUSTRY

### Problems in Transporting Vapors

Unlike oil, natural gas exists as a vapor at atmospheric temperatures and pressures. This difference calls for drastically different modes of transporting, storing, and using the two energy sources.

Oil can be shipped by truck, tanker, barge, and rail and stored in the simplest of vessels. But without costly liquefaction facilities, the movement of gas is confined to fixed pipelines. This, in itself, tends to limit the number of companies that attempt to enter any single market. (In 1982, 85 percent of all interstate gas shipments were handled by twenty pipeline companies.) Given the scale of capital required to establish a pipeline system, it is not surprising that a number of distinct firms secured footholds in separate markets during the various pipeline booms. Moreover, the costs and difficulties of gaining rights-of-way over hundreds of miles means that once a company builds a pipeline, it tends to respond to demand growth by installing (*looping*) additional pipe in the same corridor. The transnational systems of Tennessee Gas Transmission Co., Transcontinental Gas

Pipe Line Corp. (Transco), Panhandle Eastern Pipe Line Co., and Northern Natural Gas Co. (InterNorth, Inc.) all support four or five parallel pipes along their main rights-of-way.

## Isolation of Production and Distribution

Corporate linkages between gas production and distribution were even fewer during the earliest decades of the gas industry. The only company today with major activities in both gas production and distribution (Columbia Gas Systems) is centered in the Appalachian region, where natural gas was first exploited for nearby urban use.

For more than half a century, the gas-distribution business was more closely aligned with coal than with oil production. Manufactured gas depended upon coal feedstocks, and thus even in end-use markets there was no cause for antagonism between coal producers and gas distributors. Gas was, after all, used almost exclusively as an illuminant—a use for which coal was thoroughly unsuited.

Meanwhile, natural gas was mostly an incidental and often troublesome byproduct of the quest for oil. It was not until the late 1920s that technical advances allowed leaseholders in the Monroe field of Louisiana and the Panhandle field of the Midcontinent to connect their resources to urban distribution customers. Lacking transmission facilities, the only market outlets during the early days of the gas industry were carbon-black manufacturers, who located and relocated their operations to take advantage of supply isolation.

## Linking Gas Production and Distribution
## Via Interstate Pipelines

By the late 1920s, gas-field owners and manufactured-gas distribution companies both were eager to make physical connections between their facilities. Known reserves of nonassociated gas were virtually untapped, while gas produced in association with oil was mostly flared or sold for a pittance to carbon-black plants. At the same time, consumers and distributors were becoming increasingly aware of the superior burning qualities of natural gas over its manufactured cousin.

For these reasons, distributors and producers joined to organize and finance the companies that pioneered interstate gas transmission from 1925 through 1931. For example, among the companies that put together the Atlantic Seaboard pipeline were Columbia Gas & Electric (a major distributor in the East) and Standard Oil of New Jersey (a major producer). Columbia was also instrumental in formation of the Panhandle Eastern system, built in 1931. Standard Oil of New Jersey was a major participant in the Colorado Interstate pipeline and, along with People's Gas Light & Coke (Chicago's gas distributor), established the system known today as the Natural Gas Pipeline Company of America.

## THE HEYDAY OF THE HOLDING COMPANY

### An Excess of Integration

Fundamental to the structure and development of the gas industry prior to the Great Depression was the pervasiveness of *holding companies*. Through creation of a holding company, a single stockholder or stockholding group could obtain controlling interests in several (and often dozens of) distinct companies. By 1910, practically all big-city gas-distribution enterprises, except those in New York City, Baltimore, and Washington, D.C., were appendages of holding-company systems. By 1932, corporate agglomeration had reached the point that sixteen massive holding companies controlled almost half of all gas transmission and distribution in the United States. By 1935, nine holding companies controlled 80 percent of the total interstate pipeline mileage.

A good example of the scale of holding companies was Cities Service Company, circa 1930. That one business entity controlled more than sixty-five distinct utilities engaged in the transmission and sale of gas and electricity in twenty states. Cities Service also controlled forty-five companies in the oil business.

Holding companies flourished after 1888, when the state of New Jersey enacted legislation that, for the first time, empowered a company (rather than just an individual) to own stock in another firm. The effects of New Jersey's pioneering legislation are seen even today, with many U.S. firms *domiciled* in that state. While the Sherman Antitrust Act of 1890 forbade stockholders to assign voting rights to trustees (who, characteristically, were also assigned voting rights from other firms engaged in the same business), there was no law against similar affiliations of firms via outright ownership of stock by the same person or persons.

The benefits of holding company structures were multiple, but the most important were as follows:

1.  The holding-company structure provided an opportunity for growth with far less capital outlay than merger, amalgamation, or outright purchase of another company. As long as stock ownership was diffuse, a holding company could usually gain a *controlling interest* by acquiring far less than 50 percent of the *common stock*.

2.  The holding-company arrangement enhanced the ability of a single company to obtain financing because lenders would contribute money to the parent based upon the combined credit strength of the sibling enterprises. Holding companies were, in fact, the only vehicle for securing money to expand gas-distribution plant during the early part of this century. Holding companies were also uniquely suited for financing long-distance gas transmission during the first pipeline boom of the late 1920s. Lending institutions viewed these ventures as far too risky for major investment, but they were willing to contribute capital directly to the big holding companies, with little if any oversight as to how that capital was used.

3.  The holding-company form of business organization enabled gas companies to bolster system efficiencies and to wield market power through *vertical integration* (control of production, transmission, and distribution of a single commodity), *horizontal integration* (control of gas and electric utilities), and *geographic integration*.

The holding-company system, however, also had undesirable features that eventually prompted the U.S. government to dismantle it. Characteristics of holding companies that provoked public concern, and occasionally outrage, included the following:

1.  Financing was often thinly capitalized and *stock watering* was frequent. Shareholders other than the controlling party had no knowledge of the diminishing stock value because corporate structures were complex and often fluctuated from day to day. These practices eventually led to a collapse of stock values at the

onset of the Great Depression and inspired Congress to create the Securities and Exchange Commission in 1934.

2. Holding companies began to expand excessively; unwieldiness and "scatteration" superseded economies of scale. The combined management of multiple enterprises proved to be more of a liability than an asset for constituent companies and for the nation's economy as a whole.

3. State and local governments that issued franchises free of market competition and regulated local gas and electricity distributors were increasingly unable to prevent what they believed were unfair intercorporate charges when the utility was controlled by a holding company. The unregulated parent would milk its subsidiaries by entering into padded service and supply agreements that rendered ineffective any state and local supervision of cost. Because interstate gas-transmission companies were not then regulated, the holding-company structure that put gas transmission and distribution under the same management made it especially difficult for state utility commissions to protect consumers from price gouging.

State commissions were, therefore, among the most vocal proponents of the Public Utility Holding Company Act of 1935.

### The Public Utility Holding Company Act of 1935

The Public Utility Holding Company Act (PUHCA) vested the Securities and Exchange Commission (SEC) with vast powers over any company that owned or controlled 10 percent or more of the voting securities of a company engaged in generating, transmitting, or distributing electric energy or distributing natural or manufactured gas. The act empowered the SEC to approve the acquisition of securities by firms subject to its jurisdiction (*registered* companies) and gave the commission authority to scrutinize the terms of service and supply contracts that had so frustrated state and local regulatory bodies. Moreover, the act directed the SEC to supervise the restructuring and divestiture of corporations to yield "single coordinated systems confined in operations to single areas or regions . . . not so large as to impair . . . the advantages of localized management, efficient operations, and the effectiveness of regulation. . . ."

The act, in effect, called for an end to excesses in vertical, horizontal, and geographic integration. Specifically, it forbade most affiliations between electric and gas companies that were not municipally owned and whose operations crossed state lines.

In most cases, holding companies submitted their own plans of corporate divestiture for SEC approval, and many abandoned the holding company structure altogether. Although some level of concentration was still allowed, the act imposed such a high degree of oversight of company activities that over the next several decades most firms sought exemptions by *spinning off* enterprises.

In 1981, American Natural Resources Company (parent of Michigan-Wisconsin Pipeline Company and Great Lakes Gas Transmission Company) divested itself of its remaining gas distributor (Michigan Consolidated Gas Company). This left only two registered holding companies with gas-transmission and -distribution affiliates: Columbia Gas System and Consolidated Natural Gas Company. Both companies now hold mere fragments of their earlier domains.

Soon after passage of the PUHCA, Columbia Gas & Electric split into two separate companies, one gas and the other electric, and the new gas company was ordered to spin-off the Panhandle Eastern pipeline system as well. Today, companies within the Columbia fold include two interstate gas-transmission companies and seven intrastate distributors (operating in seven states).

Consolidated Natural Gas Company is only a remnant of the Standard Oil of New Jersey empire (now Exxon). The divestiture left Consolidated with its interstate gas-transmission network, plus five major distributors (including East Ohio Gas Company and Hope Natural Gas Company).

Although Cities Service Gas Company (acquired by Northwest Pipeline Company in 1982 and renamed Northwest Central Gas Company) is no longer a holding company, the old Cities Service conglomerate was one of the biggest. Its divestiture was geographic. Arkansas-Louisiana Gas Company took over the southern operations and a smaller company held onto the New England utilities. Cities Service itself retained the gas facilities in Oklahoma, Missouri, Kansas, and Nebraska.

Fifteen years after passage of the PUHCA, holding company control of interstate gas pipeline mileage had shrunk from 80 to 18 percent. New interstate pipelines, organized and built after 1935, almost always chose to avoid the act's jurisdiction by remaining completely free of distributor entanglements.

In the early 1980s, advocates of repealing the PUHCA were gaining support. They maintained that the SEC approval process for company acquisitions was cumbersome and put utilities at a disadvantage compared to nonregulated bidders. They argued, too, that the act discouraged diversification by gas distributors and that diversification could combat the credit problems that those companies currently faced.

Spokespersons for the repeal movement argued that elimination of SEC control would not signal a return to the dark ages. Since passage of the PUHCA, other regulatory mechanisms have evolved to handle the problems more specifically. Utilities are now required to keep a *unified system of accounts*, and the Natural Gas Act of 1938 instituted federal oversight of interstate gas transmission.

The opposition, nevertheless, was formidable. Even if regulation were successful in preventing questionable intercorporate transfers of funds and charges, there is no solution to the problem of shared credit strength. Quite simply, if a coal-mining affiliate of a gas distributor fails, then the bond rating of the utility itself suffers.

In 1982, New Mexico legislators passed a moratorium on utility diversification, pending completion of a study on whether and to what extent diversification ought to be regulated. Governor Anaya went one step further, proposing legislation that would prohibit utilities from entering into nonutility businesses.


## The Natural Gas Act of 1938

While the Public Utility Holding Company Act instigated the breakup of gas-industry affiliations, another piece of New Deal legislation ensured their demise. For decades, states had attempted to regulate the companies that carried gas from remote fields to local distributors, only to be admonished by the courts for meddling in matters of interstate commerce. The Natural Gas Act of 1938 (NGA) asserted federal jurisdiction over interstate gas transmission.

Companies engaged in petroleum exploration and development have traditionally been among the most free-wheeling corporate entities in America. Drilling was and still is a game of chance. Accustomed to an environment that offers big losses and big gains, oil and gas producers generally scorn the staid business of utility ventures. The antipathy increased through the years as the administering agency, the Federal Power Commission, made it clear that its interpretation of the loosely worded Natural Gas Act would keep the transmission companies on a short leash.

## THE QUEST FOR LONG-TERM STABILITY

### Life Insurance Companies and Pipeline Financing

Fortunately, at about the same time that utility holding companies were torn asunder, life insurance firms took an interest in interstate pipeline bonds. The Great Depression created a situation in which the total supply of corporate debt instruments fell short of the increases in investible life funds. At the same time, railroad bonds fell into disrepute. Gas-transmission systems were, on the other hand, becoming more attractive. Pipeline design was improving, and geophysical techniques for estimating gas reserves were gaining credibility. Moreover, states had exerted jurisdiction in the production sector to ensure sound field practices.

Northern Natural Gas Company executed the first pipeline financing with a life insurance firm. In 1935, four of the big New York insurance companies refinanced Northern's system that had been completed in 1930. They purchased twelve-year mortgage bonds at 4.75 percent interest, enabling Northern to refund its short-term renewable bank notes that carried 7.30 percent interest charges.

Heretofore, life insurance companies had purchased bonds only from "seasoned" enterprises, since they were directed by state law and practice to manage their funds conservatively. Although not seasoned in the traditional ten-year sense, Northern's system had been operating for five years, and, given the oversupply of capital in the 1930s, the transaction was not surprising. Southern Natural Gas Company and Panhandle Eastern Pipe Line Company quickly followed suit, refinancing their costly bank notes.

Insurance-company participation in pipeline bonds signalled a new era of innovative loan instruments. The lenders protected themselves against possible default by attaching novel *covenants*. These covenants often mandated regular payments into a *sinking fund*, posted *liens* on company property, restricted further issuance of bonds to prevent dangerously high *debt/equity ratios*, and required third-party assessments of gas reserves dedicated to the pipeline, along with annual updates that would automatically trigger accelerated payments if the reserves proved less than expected. Moreover, the lenders insisted on long-term (usually *minimum bill*) gas sales contracts to distributors.

By the end of World War II, and with federal regulation firmly in place, the Big Five life insurance companies felt confident enough to invest even in new pipelines owned by new firms. Nevertheless, commercial banks were (and still are) vital financiers. Bank notes are crucial to the construction phase because life insurance companies will purchase bonds only after all dangers of construction mishaps and delays have disappeared.

By 1950, the major life insurance companies held 78 percent of all interstate gas-transmission bonds, and some pipelines were financed with as much as 90 percent debt. (This 90/10 debt/equity ratio contrasts with a 1981 average for gas utility *capitalization* of 43/57.) Smaller institutional investors (like casualty insurance firms) purchased the bulk of distribution bonds, while gas exploration and production fed on company reinvestment of net income or the sale of additional shares of common stock to individuals interested in high-risk/high-return ventures.

## Long-Term Supply and Sales Contracts

The Federal Power Commission administered the Natural Gas Act's provisions for pipeline certification in a manner that echoed the concerns of lenders. The FPC required applicants to secure long-term contracts at both ends of the project. The existence of long-term contracts had an interesting side effect. Though the involved production, transmission, and distribution companies were usually not affiliated, long-term contracts promised the same supply and market security as vertical integration. This situation (though diminishing) remains today and blunts the motivation that would otherwise be compelling to establish corporate linkages among the three sectors.

The practice of wedding buyers and sellers through long-term contracts both *upstream* and *downstream* has become increasingly troublesome. In the 1970s, long-term contracts locked producers into relatively cheap sales, and distributors and direct industrial customers filed claims against interstate transmission companies who failed to deliver specified volumes. In the 1980s, the reverse was true. The

Natural Gas Policy Act of 1978 unleashed a torrent of exempt (and therefore high-cost) gas purchases. Almost every interstate transmission company was caught by surprise when higher prices, coupled with the recession, drove customers away. By late 1982, many of the major pipelines were forced to shed excess supplies in inefficient ways. They found it cheaper to accept contracted volumes of high-cost gas, like deep gas and imports, and to shut in their cheaper, more flexible supplies, than to make *deficiency payments* under *take-or-pay* terms for gas that they did not accept.

Over a ten-year period beginning in 1973, gas utilities experienced both the upside and the downside of a market system fraught with institutional rigidities. Increasingly, transmission and distribution companies are questioning the merits of security versus flexibility in supplies and sales. If gas transactions eventually shed some of these long-established rigidities and begin to resemble the crude oil trade, company managers throughout the industry might find the prospect of upstream and downstream corporate mergers increasingly attractive.

## STRUCTURAL EFFECTS OF WELLHEAD-PRICE REGULATION

### Wellhead Regulation of Affiliated Producers

Since its entry into gas pipeline regulation, the Federal Power Commission has regulated the field price of natural gas if the seller (producing company) is *affiliated* with the buyer (interstate transmission company). The commission felt that without this oversight, the sale price would be unreasonably high because the regulated arm could expect to pass on inflated charges to hungry markets, deriving greater profits for companywide operations. If, however, the two companies were not affiliated, the transmission company would likely strive for a better bargain via *arm's-length* transactions. Sales by *independent* producers—predominantly the major oil companies—were therefore free of wellhead-price regulation for many years.

This policy was a powerful deterrent to vertical integration of the gas production and transmission sectors, especially following the Supreme Court decision on the *Hope Natural Gas* case (320 U.S. 603 (1944)). The court upheld the commission's practice of computing *cost-of-service* for a regulated utility based on *original cost* rather than *replacement* or *reproduction cost*.

Following the *Hope* decision, production of gas by transmission-company affiliates dropped enormously. Transmission and distribution companies produced only 13 percent of their aggregate gas supplies in 1955, compared with 36 percent ten years earlier. Alarmed by this trend, the Federal Power Commission attempted to blunt the disincentives for affiliated gas production. In 1954 the commission began allowing affiliated producers to charge prices corresponding to a "fair field price," which brought them in line with the average price of unregulated *independent* transactions. As usual, the ruling was challenged, and in 1955 a district court reversed commission policy. The court directed the FPC to consider original costs at least "as a basis for comparison" (*City of Detroit v. FPC*, 230 F.2d 810 (1955)).

### Wellhead Regulation of Independent Producers

Not only did the judicial branch stifle the commission's attempt to reduce government meddling in field sales, but in a separate decision issued about the same time, the Supreme Court forced the commission to intrude even further. In the historic *Phillips* decision, the court ruled that wellhead price regulation must be extended to independent as well as affiliated producers engaged in interstate commerce (*Phillips Petroleum Co. v. State of Wisconsin,* 347 U.S. 672 (1954)).

It is an understatement to observe that independent producers were unhappy with the *Phillips* decision. Almost universally, economists attribute the later decline in gas reserves and production to the disincentives attendant to wellhead-price regulation. Were it not for the fact that enormous volumes of gas were already committed to pipelines under twenty- or thirty-year contracts, and for the fact that gas was often an incidental discovery in the search for oil, the shortages of the 1970s might have been far more devastating.

One other important effect of the *Phillips* decision was the fact that it put affiliated producers on a par with independents. Accordingly, interstate transmission companies began expanding once more into the production sector. Some transmission companies started their own production arms. Others entered the business by purchasing small producers or buying entire gas fields from the large independents. For example, in the five years following the *Phillips* decision:

1. Panhandle Eastern Pipe Line Company created a production subsidiary (Anadarko Production Company) that became a major part of companywide operations, placing Panhandle Eastern among the top twenty gas producers in recent years;

2. Texas Eastern Transmission Corp. purchased outright (rather than entering into a long-term supply contract) the Rayne Field in southern Louisiana for $134 million and acquired La Gloria Oil and Gas Company; and

3. Tennessee Gas Transmission Company paid a shocking $159 million for about 1 tcf of gas in a Louisiana field and created a production subsidiary. The production subsidiary (Tenneco) was so successful that the two reversed roles. Tenneco is now the parent company, and Tennessee Gas Transmission contributed only 19 percent of the 1980 companywide profits.

Outright purchase of nonassociated gas reserves and entry into the gas-production business became an attractive option for gas-transmission companies for two primary reasons. First, by owning reserves, a transmission company had complete control over deliverability. It could increase annual production as demand grew, and it could alter daily takes to correspond with seasonal swings in use. Second, by selling gas reserves to a transmission company in one lump sum, the original owner escaped wellhead-price regulation altogether—that is, of course, as long as the transaction did not seem sufficiently scandalous to prompt commission disapproval.

## The Intra- Versus Interstate Schism

Field regulation of gas that entered interstate commerce made it increasingly difficult for interstate pipelines to bid supplies away from their unregulated counterparts operating in producing states. Wellhead-price control was not, however, the only factor. The FPC's policy on *life-of-the-field* commitments was also influential.

Anticipating the *Phillips* decision, some producers had inserted clauses into their interstate contracts that terminated sales in the event that a federal agency asserted jurisdiction over the transactions. Some contracts, therefore, became void following the 1954 Supreme Court ruling.

To prevent a debilitating loss of interstate supplies, the FPC ruled that termination clauses were unenforceable because the Natural Gas Act had awarded the commission broad powers to oversee *abandonment* of gas facilities and services. The commission asserted that once gas entered interstate commerce, it was committed "for the life of the field," irrespective of the details of any particular contract. That policy survived judicial review; in 1960 the Supreme Court laid the question to rest in *Sunray Mid Continent Oil Company v. FPC* (364 U.S. 137 (1960)).

The skewed regulation of intra- and interstate gas commerce, coupled with the FPC's ruling on field commitments, powerfully affected the gas industry. Most visibly, it meant that the gas shortages of the 1970s never touched the producer states—industrial boilers in Texas consumed gas without interruption, while industries and even schools in Ohio had to shut down during the most severe winter weeks. Distorted regulation also transformed the underlying structure of the industry. Because of it, two distinct (and sometimes redundant) systems of pipelines evolved in the United States. Then, too, while oil companies shunned investment in interstate pipeline facilities, many found that state regulation was tolerable, and they bolstered their marketing options by expanding their gas *gathering systems* into pipeline utilities monitored by state commissions.

## The Natural Gas Policy Act of 1978

The Natural Gas Policy Act of 1978 (NGPA) tackled the inter-/intrastate schism—first, by subjecting *intra*state wellhead sales to the same controls as those imposed on *inter*state transactions and, later, by deregulating both. Nevertheless, intra- and interstate pipelines retained their distinct operations and ownerships because the transportation tariffs of the former are regulated by state PUCs and those of the latter are regulated by the Federal Energy Regulatory Commission. The NGPA did, however, put a damper on the perverse incentives that had long induced redundancies in the nation's transportation network.

Likewise, producers and buyers began to take an interest in optimizing transportation arrangements, utilizing both intra- and interstate facilities—a practice that was almost nonexistent during the post-*Phillips*, pre-NGPA era. In December 1978, FERC authorized single-state carriers to transport gas on a contract basis for interstate pipelines (and vice versa) without triggering federal review of transmission charges. The NGPA also put an end to abandonment restrictions on gas dedicated to interstate commerce by exempting all new gas contracts (executed after 1977) from the provisions of the 1938 Natural Gas Act.

### The Mid-Louisiana Case

In addition to eliminating the gas allocation inequities between producing and consuming states, the Natural Gas Policy Act was designed to stimulate gas exploration and development. FERC's interpretation of the act, however, boded ill for transmission companies that had previously found it attractive to invest in gas production. FERC decided that the relaxation of wellhead-price controls applied only to producing companies that were not affiliated with the purchasing transmission company.

Mid-Louisiana Gas Company launched a protest and was joined by other integrated gas companies. The plaintiffs argued that, in passing the NGPA, Congress did not intend to erase ten years of regulatory practice that drew no distinction between gas production by affiliated and independent firms. Cost-of-service treatment of affiliated gas production was abandoned in the late 1960s when the FPC enforced area rates. Again in the 1970s, producer/buyer affiliations made no difference in the application of the new nationwide ceiling prices.

In 1981, a lower court ruled in favor of the plaintiffs (*Mid-Louisiana Gas Company v. FERC* 664 F.2d 530 (5th Cir. 1981) ). On appeal, the U.S. Supreme Court ruled 5 to 4 in favor of Mid-Louisiana's position. Following that June 1983 ruling, pipeline companies could confidently invest in gas exploration and production—at least until Congress changes the rules of the game once again.

## INDUSTRY STRUCTURE IN THE POST-NGPA ERA

### The Growth of "Contract" Transportation Services

It was not until 1978 that Congress mandated a legislative solution to the nation's gas predicament. Until then, the only avenue for relief lay with the Federal Power Commission, predecessor of FERC. The commission did institute some creative programs to counteract the effects of the 1970s gas shortages. (See Chapter 5.)

The *self-help program* was one of the most successful of the FPC innovations. The commission ruled that because its jurisdiction applied only to "sales for resale," wellhead-price ceilings would not apply to gas sales directly negotiated by a producer and an industrial customer. In order for the program to work, however, interstate transmission companies had to file special tariffs for contract transportation services.

Until this time, interstate gas pipelines in the United States had functioned almost exclusively as private carriers, buying gas in the field and selling it, along with transportation services, to distributors, direct industrial customers, and to other gas pipelines. Tariffs, therefore, reflected both the price of gas and its shipment costs. Only rarely did a pipeline enter into a contract to carry gas owned by somebody else. And no gas pipeline was ever certified as a common carrier, which resembles contract-carrier status with the addition of prorationing. Prorationing ensures equal access to the facility by all interested shippers (new and old, affiliated and nonaffiliated), but its drawback is that it prevents anybody from obtaining iron-clad commitments for transporting an assured volume.

The self-help program of the 1970s may have set in motion forces that ultimately will transform the gas-transmission business. For the first time since the interstate market arose, users and producers were directly in contact. Even after the shortages disappeared, the gas surplus of the early 1980s rekindled self-help transactions. This time, industrial customers were shopping around for gas that would cost *less* (including transportation charges) than gas purchased from an interstate carrier.

Increasingly, interstate pipeline companies are having to reckon with a future that could strip them of a traditionally central and aggressive role in gas industry transactions. Some are fighting the winds of change by making it very difficult for third parties to utilize their facilities on a contract basis. This stance, coupled with the widespread belief that imprudent purchases of high-cost gas by interstate pipelines caused tremendous and debilitating rate hikes in the early 1980s, inflamed consumer-state members of Congress. In mid-1983, both houses were considering bills that would do more than simply ensure contract access to interstate pipelines. Congress was taking a serious look at imposing common-carrier status on all gas pipelines serving interstate customers.

If private-carrier status loses ground, either through a gradual shift in institutional arrangements or via congressional mandate, the loss need not be a financial one. For as long as the Natural Gas Act has been in force, pipelines have never been allowed to profit from the buying and selling of gas. Purchased-gas costs flow directly through to downstream customers; the companies earn a return only on their transportation services. Accordingly, it should make little difference to a pipeline's balance sheet whether gas that it carries is its own or belongs to somebody else.

A total transition of the U.S. pipeline network to a contract- or a common-carrier mode is not, however, free of drawbacks. Natural gas is far more difficult to store than crude oil or refined products, and it is practically impossible to move outside of fixed pipelines. Buyers and sellers do not, therefore, have the temporal or spatial latitude to err in market forecasts. But a pipeline operating as a private carrier, responsible for balancing the flow of gas from a variety of producers to a variety of individual customers, has an increased ability to smooth out the peaks and troughs of individual customer accounts.

Congress mandated common-carrier status for interstate *oil* pipelines in 1906, through passage of the Hepburn Act. Although common carrier designation had been considered in early drafts of the Natural Gas Act, the final version of the 1938 act was silent on this matter. Absent a congressional mandate, gas pipelines found it in their interest to continue conducting their operations as private carriers.

History, however, reveals several important exceptions. In 1954, Gulf Interstate Company built a pipeline from the Louisiana Gulf Coast to West Virginia, connecting with the established pipeline network of Columbia Gas Transmission Company. During its first four years, the Gulf Interstate pipeline was exclusively a contract carrier of gas owned by Columbia. In 1958, Columbia purchased the Gulf Interstate system and began managing it through Columbia Gulf Transmission Company.

Similarly, Houston Corporation (now Florida Gas Transmission Company) pioneered the Florida market in 1959 by carrying gas on a contract basis for major electric utilities that had arranged direct purchases from Texas producers. Twenty years later, about half of the load carried by Florida Gas Transmission was still handled on a contract basis.

In early 1983 FERC was in the midst of reviewing a joint application for construction of new transmission facilities in New England, coupled with increased imports of Canadian gas. Under the terms of the Boundary Gas venture, a dozen distributors in New England would purchase Canadian supplies directly at the border and contract Tennessee Gas Transmission Company to carry the gas through facilities built expressly for this purpose. Although the Energy Regulatory Administration approved the import plan in mid-1982, by the time FERC began its review, other considerations posed problems for the applicants. In addition to criticizing rigorous take-or-pay provisions extracted by Canadian sellers, opponents argued that the contract agreement between Tennessee and the Boundary distributors would place the entire burden of pipeline amortization on the purchasers—whether or not their markets would fully accommodate the anticipated volumes of committed gas.

Finally, if the Alaska Natural Gas Transportation System (ANGTS) is built as presently certificated, it too will function as a contract carrier in both its Alaskan and Canadian sections. Canada provides perhaps the best examples of contract gas carriers. The intraprovincial pipeline in Alberta (formerly Alberta Gas Trunkline, now NOVA), for instance, operates exclusively on a contract basis. The major west-east carrier, TransCanada Pipeline Company, also moves a great deal of gas owned by other entities.

### The Pipeline Network Matures

Diminishing opportunities for new pipeline construction are far more threatening to a company's balance sheet than changes in its carrier status. Absent new investments in gas plant, a pipeline company can expect its profits to fall in accordance with its *vanishing rate base* because federal regulation allows earnings only on the *original costs* of facility construction (sometimes called *historic costs*) that have not yet been recovered through the *depreciation* component of a pipeline's tariff.

Pipeline expansion depends upon continuing growth in gas markets or the replacement of old and corroded equipment with new investment. The drive to replenish rate base is so strong that the industry is replete with folklore of company decisions to abandon and replace perfectly serviceable gas plant simply because the facilities were fully depreciated and no longer earning a return.

If market growth is limited because of stagnation or decline in either supplies or demand, the outlook for new pipeline investment is bleak. In the 1970s, diminishing supplies of gas severely curtailed pipeline construction—except for the storage projects and supplemental-gas ventures discussed in Chapter 4. In the early 1980s, conditions reversed, and consumer reaction to higher prices restricted market expansion.

Unless the outlook for pipeline construction improves or the transmission industry is deregulated, gas-transmission companies will either have to continue investing in existing facilities by replacing debt capital with equity—a rather inefficient use of investible funds—or they will have to look for investments in other lines of business. Especially following the favorable 1983 Supreme Court ruling in the *Mid-Louisiana* case, investment in gas production became no less attractive to gas utilities than it was to independent producers. But downstream investment in gas-distribution companies offers few opportunities because of the same market limitations that now restrict pipeline growth. Distributor integration also bears the disadvantage of greater regulation through the Public Utility Holding Company Act.

Finally, from a practical standpoint, distribution-company investment in the early 1980s is not very attractive. The financial health of local gas companies is weak in general and in some cases, desperate. The average return on equity for distributors dropped from 15.5 percent in 1979 to 11.0 percent in 1981. This contrasts with an improvement in gas pipeline returns over the same period, from 16.0 to 17.7 percent.

## Spin-Offs of Distribution Affiliates

Relatively few interstate transmission companies now support distribution affiliates, and in 1981, two of them announced plans to sever their remaining ties. The Natural Gas Pipeline Company of America (now a subsidiary of MidCon Corporation, which also owns oil pipelines) divested itself of one of the oldest distributors in the United States. Peoples Gas Light & Coke of Chicago had, in fact, been the biggest investor in the infant transmission company. Meanwhile, American Natural Resources Company, parent of Michigan-Wisconsin Pipe Line Company, spun off Michigan Consolidated Gas Company, which is one of the nation's biggest distributors. Six years earlier, ANR had spun off another leading distributor: Wisconsin Gas Company.

Statements made by the two transmission firms indicate that the comparatively low earnings of the distribution affiliates led to the company shake-ups. Peoples Gas Light & Coke contributed 41 percent of companywide revenues in 1980 but only 25 percent of net income. Worse yet, Michigan Consolidated accounted for 47 percent of its parent's revenues but only 11 percent of net income. Its balance sheet was so poor that *Moody's Public Utilities* observed in 1980 that:

> Michigan Consolidated Gas Company . . . again reported wholly unsatisfactory earnings. While greater than in 1979, net income in 1980 was still less than one-third of that authorized by the Michigan Public Service Commission. Like its pipeline affiliate, Michigan Consolidated suffered greatly reduced sales. In addition, the uses of historical costs and the long delays involved in the present rate-making process have made it impossible for the company to achieve a reasonable return on its investment (Moody's 1982).

### Joint Ventures among Transmission Companies

By the early 1980s, the structural linkages between transmission and distribution companies were sorely strained. Transmission companies were, however, strengthening their ties with one another—a movement that had its roots in the gas shortages of the 1970s. The 1970s gave birth to a number of supplemental-gas projects, most of which were *joint ventures* sponsored by two or more transmission firms. The extraordinary capital outlays required for LNG terminals, coal-gasification plants, and Arctic pipelines compelled gas-transmission companies to pool their financial strengths. Joint ventures were also popular for major storage projects and offshore pipeline systems. Table 8–7 lists some of these projects, along with their constituent companies.

The Alaska Highway Gas Pipeline project (with 1982 cost estimates exceeding $30 billion) is the supreme example of a joint venture. At its peak sponsorship, the partnership included a half dozen U.S. pipeline companies, three Canadian gas utilities, and the three major oil companies with gas interests in Arctic Alaska. Even with combined

**Table 8-7.**   Joint-Venture Gas Projects.

| Project | Sponsors |
|---|---|
| *Completed or Under Construction in 1983:* | |
| High Island Offshore System (HIOS) 200 miles of pipeline in the Gulf of Mexico | Michigan-Wisconsin Pipeline (operator) United Gas Pipeline Co. Texas Gas Transmission Corp. Transcontinental Gas Pipeline Co. Natural Gas Pipeline Co. of America |
| Sea Robin Pipeline Company 400 miles of pipeline in the Gulf of Mexico | United Gas Pipeline Co. (operator) Southern Natural Gas Co. |
| Bear Creek Storage 32 bcf of underground storage in N. Louisiana | Tennessee Gas Transmission Co. Southern Natural Gas Co. |
| Algeria I LNG LNG import terminals in Maryland and Georgia | El Paso Natural Gas Co. Columbia Gas Transmission Co. Consolidated Gas Supply Corp. Southern Natural Gas Co. |
| Trailblazer Pipeline 800 miles of pipeline from Wyoming to Nebraska | Natural Gas Pipeline Co. (operator) Columbia Gas Transmission Co. InterNorth, Inc. (Northern Natural) Tennessee Gas Transmission Co. Colorado Interstate Gas Corp. Mountain Fuel Resources, Inc. |
| Ozark Gas Transmission System 300-mile pipeline to tap gas in the Arkoma Basin | Texas Oil & Gas Corp. (operator) Tennessee Gas Transmission Co. Columbia Gas Transmission Co. |
| Northern Border Pipeline 800 miles of pipeline from the Manitoba border southeastward into the United States | InterNorth, Inc. (operator) Panhandle Eastern Pipeline Co. United Gas Pipeline Co. Northwest Pipeline Co. TransCanada Pipelines (of Canada) |
| "Prebuild" (Canadian connection with Northern Border) | Nova Corp. Westcoast Transmission Co. TransCanada Pipeline Co. Alberta Natural Gas Co. |

**Table 8-7.** *(Continued).*

| Project | Sponsors |
| --- | --- |
| Great Plains Coal Gasification 125 mmcf/day plant in North Dakota | American Natural Resources (operator) (and parent of Michigan-Wisconsin) Tennessee Gas Transmission Co. Transcontinental Gas Pipe Line Co. Columbia Gas Transmission Co. Pacific Lighting Corp. |

*Speculative Ventures (as of 1983)*

| Project | Sponsors |
| --- | --- |
| Rocky Mountain Pipeline 600-mile pipeline from Wyoming to California | Pacific Gas & Electric Co. Pacific Lighting Corp. El Paso Natural Gas Co. Northwest Pipeline Co. |
| Trans-Anadarko Pipeline pipeline connection from north Texas to Louisiana | United Gas Pipeline Co. Texas Gas Transmission Corp. Tennessee Gas Transmission Co. Southern Natural Gas Co. |
| Western LNG LNG import terminal in southern California | Pacific Gas & Electric Co. Pacific Lighting Corp. |
| Alaska Natural Gas Transportation System (Alaska segment) | Northwest Pipeline Co. (operator) Pacific Gas & Electric Co. Pacific Lighting Corp. InterNorth, Inc. (Northern Natural) Panhandle Eastern Pipe Line Co. United Gas Pipeline Co. Columbia Gas Transmission Co. Texas Eastern Transmission Corp. Texas Gas Transmission Corp. (withdrawn) Michigan-Wisconsin Pipe Line Co. (withdrawn) Exxon Atlantic Richfield Co. Standard Oil of Ohio |

pipeline and producer support, the consortium repeatedly failed to attract the requisite debt capital. In the spring of 1982, the sponsors announced an extended delay.

Joint ventures became a popular form of financing for conventional projects as well. Following passage of the Natural Gas Policy Act, several joint ventures pioneered new pipeline routes to tap burgeoning supplies of unregulated deep gas and Canadian imports. Five transmission companies formed Northern Border Pipeline Company (also known as the "Eastern Leg" of the proposed Alaska Natural Gas Transportation System), which commenced shipments of Canadian gas through its 823-mile facility in September 1982. Six companies contributed equity to the $550-million Trailblazer pipeline. Trailblazer provided a vital link between deep-gas producers in the Rocky Mountain Overthrust Belt and the big trunklines that transect central Kansas. In late 1982, four interstate transmission companies pressed for prompt certification of their planned Trans-Anadarko Pipeline System, spanning the region between deep-gas fields in Oklahoma and trunkline facilities in northern Louisiana.

## Mergers and Corporate Takeovers

The gas industry entered a chaotic period in the early 1980s, which made it possible for one company to see potential in what another viewed as disaster. In 1982, amidst a nationwide merger mania, Northwest Gas Pipeline Company bought a controlling interest in one of the oldest (but most financially stressed) gas companies in North America. The utility properties of Cities Service Gas Company (now Northwest Central Gas Company) spanned six states in the central region of the United States. The transfer of the company's utility plant came on the heels of Occidental Petroleum Company's $4-billion purchase of Cities Service's oil and gas production assets. Northwest's role in the takeover was an ironic twist. In its infancy, Northwest had been bought out by the dominant gas pipeline operating west of the Mississippi, El Paso Gas Company. Northwest exists today only because the courts ordered divestiture on antitrust grounds.

The break-up of producing and pipeline interests in Cities Service Gas Company was not an isolated event. Financially crushed by collapse of its LNG ventures (see Chapter 4), costly take-or-pay purchase commitments, and a shrinking gas market in California, El

Paso Gas Company was acquired by Burlington Northern, Inc., a conglomerate with large holdings in railroads and timber. Another railroad company, CSX Corporation, acquired Texas Gas Transmission Corporation, while Goodyear Tire and Rubber Company absorbed Celeron Corporation, parent of several intra- and interstate pipelines operating out of Louisiana. Tesoro Petroleum sought control of Enserch Corp., parent of Lone Star Gas, a major intrastate pipeline; and a number of other energy companies. In August 1983, Northwest Energy Corporation (parent of Northwest Pipeline Company and Northwest Central Pipeline Company) sensed that an unfriendly takeover was in the wind and sought a more compatible buyer—Allen and Company, Inc., a New York investment banking firm. In the mid-1980s, divestitures, mergers, and takeovers may well become one of the biggest issues within the industry.

### Who Is the Gas Industry?

Since its birth in the 1820s, the gas industry has cycled through various phases of integration and disintegration among its three component enterprises. Today, the industry is strikingly disintegrated. Not only are distinct companies operating as producers, shippers, and distributors of gas, but even within those three sectors, noteworthy schisms exist. Deep-gas producers have entirely different interests from producers of nonexempt categories of gas. Intrastate pipelines still operate in a regulatory climate foreign to their interstate counterparts.

These diverse interests have made it increasingly difficult for Congress to find political solutions to the nation's energy woes. There is no voice that can speak for the gas industry as a whole. What is more, history demonstrates that laws and federal programs intended to accomplish specific goals have, in addition, profoundly influenced the very structure of the gas industry and its institutions. In turn, the industry's structure and institutions have determined the efficiencies and the incentives of the gas trade and the effectiveness of regulation. Decisionmakers would be wise to anticipate likely structural effects before embarking on new legislative and administrative programs.

### REFERENCES

Moody's Investors Service. *Moody's Public Utility Manual, 1982.* New York: Moody's Investors Service, 1982.

# 9 OUTLOOK: THE RISE AND FALL OF REGULATION IN THE NATURAL GAS INDUSTRY

Readers who have followed the course of the previous eight chapters will agree that the history, structure, and economics of the gas industry are exceptionally complex. At the root of this complexity is a heritage of regulation affecting all aspects of the industry, including incentives to produce, the geography of the transmission network, and fuel choices made by consumers. Even in hindsight, it is difficult to judge whether any particular law, rule, or phase of U.S. regulatory history did a better job in promoting equity and efficiency—the twin goals of regulation—than an unregulated market could have achieved. Clearly, however, some outlived laws and rules have caused more problems than they resolved.

## EFFICIENCY AND EQUITY IN GAS REGULATION

In the years immediately following passage of the 1938 Natural Gas Act (NGA), almost no one doubted that the act would make a contribution to both equity and efficiency. By subjecting interstate pipelines to federal rate regulation, the NGA could prevent transmission companies connecting new gas fields from reaping monopoly profits. Moreover, *cost-of-service* ratemaking standards for gas transmission ensured that the *economic surplus* (the difference between market value and actual cost) could be passed on to the consumer.

From an efficiency standpoint, too, the 1938 act seemed to offer unquestioned benefits. Although economists of all schools condemned other New Deal legislation of the period that limited entry into the airline and trucking industries, entry restrictions in the gas-transmission business received nearly universal approval on the ground that economies of scale in pipeline transportation made it a "natural monopoly." In such a situation, free entry could be expected to dissipate any economic surplus as surely as an unregulated monopoly or monopsony position would transfer the surplus to the pipeline owners at the expense of consumers and producers. By issuing a limited number of permits to engage in interstate gas transmission, the government could prevent overbuilding and costly duplication of facilities.

But almost as soon as the ink had dried on the NGA, one regulatory gap after another began to appear. First was the problem of potentially abusive sales transactions between affiliated producers and pipelines. The Federal Power Commission contended that authority over these sales was essential to its mission of ensuring just and reasonable pipeline charges, and the courts subsequently upheld that claim. A decade and a half of debate ensued as to whether the FPC should, or was obliged to, extend its jurisdiction to gas sales by independent producers. In 1954 the Supreme Court settled the question, directing the commission to do what it had previously avoided, and wellhead pricing of all interstate gas "sales for resale" came under FPC jurisdiction.

During the next twenty years, the commission made several false starts in trying to craft a system of wellhead price controls that was both manageable and acceptable to the courts. Meanwhile, the inevitable occurred: By locking in prices established in an era of excess supplies and perpetuating those prices after the surplus had vanished, FPC regulation of wellhead prices created a nationwide shortage. The contrast between the rules governing intrastate and interstate gas markets concentrated this shortage in the gas-importing states, as utilities and industrial consumers in the producing states outbid the interstate pipelines for new supplies.

When oil prices shot up in the mid-1970s, gas demand followed the same course, as gas became increasingly attractive in comparison. Price deregulation seemed politically unthinkable under the circumstances, so apportionment of supply became the central gas-policy issue of the 1970s. Federal and state governments extended the reach of regulation to gas consumption, first through *curtailment schedules* and ultimately through direct controls on the end-uses of natural gas.

Congress finally addressed depressed exploration and production incentives in the Natural Gas Policy Act of 1978 (NGPA). Although the law undoubtedly encouraged industry to find and develop new gas reserves, its intricate pricing scheme for various categories of gas wells and gas producers spawned a whole new generation of administrative and legal complexities and, more important, economic inefficiencies. The outstanding example of the latter was the *deep-gas* boom of 1978–82. Exploration companies targeted strata 15,000 feet or deeper in order to qualify for exemption from price regulation under section 107 of NGPA. Anything shallower would remain under controls until 1985, when all new gas discoveries would be deregulated. Because drilling costs increase roughly in proportion to the square of well depth, the NGPA tilted exploration incentives in favor of exceptionally costly gas and away from prospects where more gas could have been found and developed for the same amount of effort and expense.

The early 1980s revealed both the unfairness and economic irrationality of the deep-gas incentives. Producers who went after these costly deposits (and the institutions that financed them) watched as the market for newly discovered gas crashed from a range of $7 to $10 per thousand cubic feet to $3 or below—less, in fact, than the price ceiling for regulated gas. Pipelines exercised "market-out" clauses wherever they could find them in existing contracts. Even absent such provisions, the pipelines were pressing producers to renegotiate both price and minimum-purchase terms. In some instances, the transmission companies brazenly refused to make deficiency payments demanded by their contracts for gas they failed to take, or unilaterally declared that subsequent takes would depend upon producer acceptance of a lower price.

Producer bankruptcies were inevitable unless transmission companies honored their purchase obligations, yet pipelines were not about to pay for gas that downstream consumers refused to take. Average gas prices nevertheless continued to rise, driven by escalation schedules in NGPA ceiling prices and in contracts for decontrolled gas, at the very time that oil prices were trending downward and the United States was in a severe economic slump. By mid-1983 it had become obvious that final markets did not contain enough revenue to meet everyone's contractual and legal expectations.

In this circumstance, widespread abrogation of high-cost gas contracts seemed necessary to the financial health—or even survival—of transmission and distribution companies. The Reagan administration actually proposed a law permitting abrogation of all gas-purchase contracts. But how could unilateral or legislative repudiation of purchase commitments be fair to gas producers? Deep-gas drillers had, after all, done exactly what Congress had intended in the NGPA. And it was the pipeline companies themselves that established prices of $6 to $10 per mcf by their frenzied bidding for deregulated gas.

At any rate, the sum of the wellhead gas prices and gas-import prices sanctioned by existing contracts and law, the interstate gas-transportation markups deemed "just and reasonable" by FERC, and the distribution markups approved by state utility commissions came to exceed the burner-tip value of gas. Somebody was bound to be hurt: Gas producers, pipelines, and distribution companies could try to impose higher prices on consumers, according to the terms of their "lawful" contracts, tariffs, and rates. But try as the companies might, consumers would not yield any more revenue. In mid-1983, the desperate attempts by each sector of the industry to recover rising costs and also to obtain its allowed return on investment was threatening to plunge them all into a *death spiral*—the self-perpetuating collapse in demand.

## THERAPEUTIC ASPECTS OF MARKET DISTRESS

Widespread distress can be expected so long as the gas industry as a whole tries to extract more revenue from the market than consumers are willing to give them. As prices go up, large industrial consumers switch to alternative fuels, and commercial and residential gas users conserve to a degree hardly dreamt of in the 1970s. In mid-1983, existing legal and regulatory arrangements were distributing the financial distress almost randomly across the industry and the general economy: High gas prices had already caused some industrial consumers to shut down operations, and many others were operating at a loss. A number of producers (and their supporting banks) had failed. The outlook was no brighter for the most distressed gas pipelines and distribution utilities, for which bankruptcy (or at least a desperate merger or reorganization) loomed on the horizon.

While the gas industry was experiencing an extraordinary level of market disorder, stress was not altogether absent from other segments of the economy and was nearly universal in the energy industries. A turnaround in oil prices, coupled with the deep recession, had caught almost everybody by surprise. Electricity and coal contracts that industrial users had locked in on a long-term basis at fixed or escalating prices had become financial liabilities. "Take-or-pay" distress was most apparent in the gas industry—but it was by no means restricted to gas utilities.

Not surprisingly, each of the producer factions, the inter- and intrastate pipelines, distributors, and the various classes of gas consumers began to ask the FERC, the Congress, the state commissions, and the courts not only to reaffirm their own legal and contractual rights but to abrogate the rights of others wherever these rights frustrated realization of their own business expectations. There was no way all of these parties could be satisfied. The regulatory and legal changes various factions urged on FERC or Congress were directed mainly at shifting the inevitable pain and suffering and the threat of bankruptcy to someone else. In the short term, at least, the benefits of almost every regulatory "fix" proposed would have a zero or even a negative sum.

In the longer term, the stress that all sections of the gas industry feel today may be more therapeutic than their purported cures and, indeed, more therapeutic than hurtful. The present combination of rising prices and surplus deliverability is fostering a host of healthy (and in many cases essential) structural reforms. For example:

1. It is keeping the pressure on all classes of consumers to invest in energy conservation.
2. It is reaffirming for industrial consumers the advantages of being able to switch fuels in response to changes in relative prices and availability.
3. It is bringing a spot commodity market into existence for natural gas and will, within a couple years, lead to formation of a futures market in which producers and industrial consumers can hedge against unforeseen price changes.
4. It is impelling gas distributors and state regulatory commissions to implement more flexible, economically realistic, and efficiency-fostering rate structures.

5.  It is forcing gas-transmission companies to reform gas-purchasing practices and to bargain for greater realism and flexibility in their present gas-purchase commitments.
6.  It is pressing the transmission and distribution sectors and their respective regulators to move toward decoupling rates for those industrial gas markets that are sensitive to alternate-fuel prices and the business cycle from residential-commercial rates.
7.  It is forcing cost-consciousness on the transmission and distribution sectors for the first time, and teaching all parts of the industry to give first consideration to the underlying economic soundness of proposed investments, rather than merely their legal merits and regulatory acceptability.

Finally, and most important in the long run, the current distress is infusing the regulated sectors of the industry with a spirit of enterprise, ingenuity, and experimentation that was not in evidence when regulation guaranteed them an unlimited market. The remarkable progress achieved since 1981 in reforming the structure, behavior, and indeed the thinking of the natural gas industry has been exceptional, and it has been due almost entirely to the real or anticipated pain and suffering that stems from a body of regulation and contracts which sanctifies rising prices in the face of falling consumption.

## UNRAVELLING THE REGULATORY WEB

There is probably no set of policies that federal regulators or Congress could devise that would move the industry so rapidly in the direction it must go as it is now moving under the spur of economic necessity. Congressmen and regulators and, indeed, most gas-industry executives tend to see the situation differently. Many of them do not really trust market pricing and decentralized decisionmaking, and prefer a visible hand, however palsied, to an invisible one. There is a widespread clamor for Congress to "do something." The question of whether to relax or once more tighten wellhead-price regulation is, of course, part of the legislative agenda. Under either alternative, however, the sum of all prices permitted by law would still exceed the market value of gas. Thus, decontrol of "old-gas" contracts, a window for renegotiation of price terms in all existing contracts, or a new "cap" on gas prices tied to the rate of inflation, would change the distribution of revenues among classes of gas producers and shift losses one way or another between the producing and utility sectors of the industry, but none of them would have much, if any, effect on final consumer prices.

The chronic gas shortage and the economic surplus formerly perpetuated by wellhead price controls are both gone, and they are gone for good. Henceforth, the price of gas will track its value. And if exploration incentives can be maintained (with the help of producer-state prorationing) even during the occasional supply gluts that are inevitable, shortages should not pose severe problems. Unless Congress overreacts to the present turmoil by reimposing price controls more strictly than any major faction has yet proposed, prices will have the flexibility to ensure that no downturn in gas supplies or unexpected upswing in gas demand will create a lasting market imbalance. And price flexibility will make it possible for transmission and distribution utilities to dispense with curtailment schedules other than for management of exceptional and unforeseeable emergencies.

The regulatory frontier in the mid-1980s will therefore be in the pipeline sector, rather than at the wellhead. In 1983, debate centered on the institutional status of interstate gas pipelines. Should pipelines cease being primarily buyers and sellers of gas, and be forced to carry gas for producers, distribution companies, or industrial consumers on a contract basis? Should restructuring of the industry go even further, imposing common-carrier status on the transmission companies?

If Congress chooses to mandate action on these questions, it is likely to work more harm than good. Every pipeline already has a transportation tariff on file with FERC, and every pipeline moves some gas on contract. Most of them have come to recognize that it is better to carry gas owned by producers or consumers than to lose their business entirely. However, it is probably in nobody's best interest for any interstate pipeline to abandon its historic role entirely as buyer and seller of gas, broker, and manager of gas inventories. Unlike oil, customers cannot store gas on their premises. Except for a minority of large industrial consumers and perhaps a few large distribution companies that have adequate underground storage, purchases from pipelines will continue to offer the most economical means of balancing the daily, seasonal, and cyclical swings in gas demand.

If Congress reacts to the clamor to "do something!" about today's transitory problems by legislating sweeping institutional changes, the new laws will probably look as mischievous in hindsight as the 1978

Natural Gas Policy Act now appears. And if Congress and FERC now try to bail the utility sectors of the gas industry out of their present embarrassment through a host of new (albeit temporary) regulatory prescriptions and proscriptions, they will only reinforce the companies' old image of themselves as wards of the state. The transmission companies will have just one more justification for making investment and operating decisions more on the basis of legal stricture and agency approval than on the basis of business judgment and common-sense.

It is time for the public and Congress to recognize that the original rationales for the economic regulation of gas production and transmission have lost whatever force they once had. There is no longer much of an economic surplus for regulation to preserve or channel to consumers. The existence of a mature pipeline network, moreover, means that transmission companies in most regions of the country are capable of interacting in a competitive fashion. Perhaps the aim of Congress should be to restructure the governing framework so that the combination of law and market transactions in the gas industry begins to resemble the functioning of natural systems. The laws of physics and biology are, after all, relatively simple, but the manifestations of those laws in weather or living organisms, for example, are stunningly complex—and self-sustaining.

Specifically, the government could get out of the business of saying anything about gas production. It would neither continue to regulate the prices for new gas discoveries—deep, shallow, tight, or whatever—nor would it break the bonds of existing contracts for old gas. By the same token, it could get out of the business of utility-type regulation of gas transportation entirely, limiting its intervention to health, safety, and environmental standards and the kind of economic surveillance—antitrust and antidiscrimination standards, for example, to which most of the economy is subject.

The federal presence in the gas-transmission business might be modeled after the Texas law governing intrastate pipeline companies that carry about one-fifth of the natural gas sold in the United States and that compete vigorously with one another both in the field and for wholesale and retail customers. Texas maintains conventional utility-type regulation of service to residential and small commercial customers, but for large-volume sales state law generally deems rates or terms of service to be "just and reasonable" if:

(1)   Neither the gas utility nor the customer had an unfair advantage during the negotiations; or

(2)   The rates are substantially the same as rates between the gas utility and two or more such customers under the same or similar terms of service; or

(3)   Competition does or did exist either with another gas utility, another supplier of natural gas, or with a supplier of an alternative form of energy (*Texas Revised Civil Statutes,* 1446c:38(b) (1978) ).

Considerable movement is possible in this direction even without new legislation. In its 1944 decision *FPC v. Hope Natural Gas Co.*, (320 U.S. 660), the Supreme Court declared that the Natural Gas Act does not dictate the methods and rules by which the FPC (and by implication, FERC today) is to regulate interstate commerce in natural gas, so long as it makes a reasonable attempt to accomplish the Act's objectives. Since that time, the Courts have steadily expanded the scope of the FPC/FERC authority, but except in the 1954 *Phillips* case, the major decisions in this direction have all been in support of the commission's own desire to regulate more and in greater detail.

The current crisis has already triggered a massive withdrawal of federal regulatory authority from the gas-transmission business, and without any change in statute. FERC rulings have moved back toward the original spirit of the Natural Gas Act, which exempts from regulation the prices pipelines charge in their "direct" (not-for-resale) sales, by approving new industrial sales programs involving tariffs that permit market pricing and frequent changes without new rate proceedings before the commission.

FERC is also encouraging pipelines and their large customers to undertake rate "settlements" outside of the usual adversary proceedings and has issued rules granting pipelines "blanket authority" for short-term *off-system* sales to other pipelines, either inter- or intrastate, without the need for commission approval of individual transactions. The benefits of flexible pricing and off-system sales, however, will probably prove greatest when the present surplus is gone. Exchanges of gas between pipelines can freely supplement actual physical shipments in optimizing the operational efficiency of the national gas-transportation system.

With the end to any persistent differential between wellhead prices in intra- and interstate sales, the transmission sector will function as an integrated national (or continental) grid, into which gas from any producing region of North America can be delivered to any consuming region—by exchange and displacement if not physically—and at a very low cost relative the final value of gas. No longer will it be possible for acute shortages and deep curtailments to exist on one pipeline system, as occurred in the mid-1970s, while a few miles away boilers capable of operating on heavy fuel oil or coal continue to burn gas. Deregulation of wellhead prices and sales among pipelines will also combine to end the large disparities in end-use prices offered by neighboring distribution companies served by different pipelines.

Much of this book traced the process by which regulation in one sphere of gas-industry activity created a situation that demanded additional government intervention. In the industry's maturity the process seems to work in reverse, as it did for the airlines, the railroads, motor trucking, and the securities business. A little bit of deregulation can be a powerful, subversive, and self-perpetuating force. Deregulation of deep gas in the 1978 Natural Gas Policy Act meant *de facto* decontrol of final consumer prices and, consequently, of average wellhead prices, as pipelines bid up the price of exempt gas so high that final consumer prices reached (and quickly exceeded) the market value of gas. One way or another, the interaction of supply and demand will now force prices down until the market clears, whether or not Congress passes new legislation on natural gas. A trend toward pipeline deregulation seems as inevitable today as the advance of regulation appeared until about 1978. Moreover, the proposal for outright withdrawal of the federal regulatory presence in gas transmission is not altogether new. As Paul McAvoy and Robert Pindyck urged in 1975:

> [T]he [Federal Power] Commission should consider relaxing, rather than tightening, its supervision of pipeline prices and profits. Our findings suggest that the pipelines face a certain amount of competition in many of their regulated markets. Such competition limits the extent to which they can raise prices above costs. That fact, together with the near impossibility of regulating prices in such a way as to eliminate monopoly profit, makes it most unlikely that regulation provides benefits worth its administrative cost. The commission should consider abandoning the cost-of-service price setting method and instead investigate the extent to which various pipeline markets are competitive. Where competition exists—where there are more

than two gas pipelines and there are close substitutes of sources of energy—it could deregulate. Where effective competition does not exist, it could set prices based upon costs and prices in the more competitive markets. Such a "comparative" price-setting technique, though approximate, is likely to be as effective as present rate-setting methods and to require less supervision. (McAvoy and Pindyck 1975: 133)

Useless laws can sometimes be ignored rather than repealed, but a deliberate dismantling of utility-type regulation affecting interstate gas transmission offers substantial advances in social efficiency, as alluded to in the previous citation. Specifically, repeal of the Natural Gas Act would eliminate the perverse incentives utility economists have long regarded as inherent and evil side effects of such regulation. For example, the threat of a vanishing rate base would no longer be a driving force in company decisions to build new plant. Rather, deregulated transmission firms would find it beneficial to continue to expand service with as little new plant investment as possible. There would no longer be any need for public servants to scrutinize company plans and accounts in order to root out redundancies, padded demand estimates, or goldplating.

And once the vanishing rate base is conquered, why should any transmission executive fret over the fact that there is precious little opportunity for new utility-plant expansion? For the first time since 1938, gas pipelines would operate from the same efficiency-fostering motives as other businesses: The less they spend on upgrading their facilities, the more profitable operations would become through time.  Finding a place to invest pipeline earnings outside of the transmission business would pose no special problem because no matter how much longer the Public Utility Holding Company Act may survive, a deregulated gas pipeline would be free to invest its earnings wherever they appeared likely to be the most productive.

Perhaps the most important outcome of pipeline deregulation would be a tendency for the various segments of a now fragmented industry to see themselves as intimately connected and mutually dependent upon the continued health of upstream and downstream participants. Gone already are the days when producers and pipelines could take gas marketability for granted and treat take-or-pay commitments and consumer guarantees as more important than the economic sense of the project itself.

Free enterprise in the gas business will not emerge overnight, but it is coming faster than anyone anticipated a few years ago, propelled by an economic logic that demands—and teaches—the benefits of diversity and flexibility in pricing, contract terms, and industrial organization. It also demands radically different ways of thinking: less legalism and less concern with regulatory and political ephemera, and a new attention to the economic and financial fundamentals that dictate success or failure in unregulated industries.

### REFERENCES

MacAvoy, Paul W., and Robert S. Pindyck. *Price Controls and the Natural Gas Shortage*. Washington, D.C.: American Enterprise Institute, 1975.

# SELECTED BIBLIOGRAPHY

## CHAPTER 2

American Gas Association. *Gas Rate Fundamentals*, third edition. Arlington, Va.: American Gas Association, 1978.

Clark, James A. *The Chronological History of the Petroleum and Natural Gas Industries*. Houston, Tex.: Clark Book Company, Inc., 1963.

"Diary of an Industry: The American Gas-Light Journal, Highlights of 1859." Dallas, Tex.: Copyright Energy Publications, Division of Harcourt, Brace, Jovanovich, undated.

Hilt, Luis. "Chronology of the Gas Industry." *American Gas Journal* (May 1950).

Lom, W.L., and A.F. Williams. *Substitute Natural Gas Manufacture and Properties*. New York: Halsted, 1976.

Peebles, Malcolm, W.H. *Evolution of the Gas Industry*. New York: New York University, 1980.

Rice, Wallace. "75 Years of Gas Service In Chicago." Chicago: The Peoples Gas Light and Coke Company, 1925.

Stotz, Louis, and Alexander Jamison. *History of the Gas Industry*. New York: Press of Stettiner Brothers, 1938.

## CHAPTER 3

American Gas Association. *Gas Facts, 1981*. Arlington, Va.: American Gas Association, 1982.

American Gas Association. *Gas Rate Fundamentals*, third edition. Arlington, Va.: American Gas Association, 1978.

American Petroleum Institute. *History of Petroleum Engineering*. Dallas, Tex.: American Petroleum Institute, 1961.

Bullard, Fredd Jean. *Mexico's Natural Gas: The Beginning of an Industry*. Austin: University of Texas, 1968.

Clark, James A. *The Chronological History of the Petroleum and Natural Gas Industries*. Houston, Tex.: Clark Book Company, Inc., 1963.

*Diary of an Industry: The American Gas-Light Journal Highlights of 1859*. Dallas, Tex.: Copyright Energy Publications, division of Harcourt, Brace, Jovanovich, undated.

Harding, Richard W., ed. *Natural Gas Distribution*. University Park, PA.: Pennsylvania State University, 1963.

Hilt, Luis. "Chronology of the Gas Industry," *American Gas Journal*, (May 1950).

Interstate Oil Compact Commission. *A Study of Conservation of Oil and Gas in the United States*. Oklahoma City, Okla.: IOCC, 1964.

Kilbourne, William. *Pipeline*. Toronto: Clarke, Irwin, and Company, Ltd., 1970.

Leston, Alfred M., John A. Chrichton, and John C. Jacobs. *The Dynamic Natural Gas Industry*. Norman: University of Oklahoma Press, 1963.

Moody's Investors Service. *Moody's Public Utility Manual, 1982*. New York: Moody's Investors Service, 1982.

Nehring, Richard (of the Rand Corporation, prepared for the U.S. Geological Survey). *The Discovery of Significant Oil and Gas Fields in the United States*. Washington, D.C.: U.S. Government Printing Office, January 1981.

Nevner, Edward J. *The Natural Gas Industry: Monopoly and Competition in Field Markets*. Norman: University of Oklahoma Press, 1960.

Peebles, Malcolm W.H. *Evolution of the Gas Industry*. New York: New York University, 1980.

Stotz, Louis, and Alexander Jamison. *History of the Gas Industry*. New York: Press of Stettiner Brothers, 1938.

Tiratsoo, E.N. *Natural Gas: A Study*. New York: Plenum Press, 1967.

## CHAPTER 4

Barlow, Connie C., (for the Alaska Department of Natural Resources). *Natural Gas Conditioning and Pipeline Design: A Technical Primer for Non-technicians with Special Reference to the Alaska Highway Gas Pipeline*. Juneau: State of Alaska, March 1980.

Peacock, Donald. *People, Peregrines and Arctic Pipelines.* Vancouver, B.C.: J.J. Douglas Ltd., 1977.

Peebles, Malcolm W.H. *Evolution of the Gas Industry.* New York: New York University, 1980.

Tiratsoo, E.N. *Natural Gas*, third edition. Houston, Tex.: 1980.

Tussing, Arlon R., and Connie C. Barlow (for the U.S. Department of Energy). *Supplemental Gas Marketing and Financing Issues.* Washington, D.C.: U.S. Government Printing Office, October 1978.

Tussing, Arlon R., and Connie C. Barlow (for the Alaska State Legislature). *An Introduction to the Gas Industry with Special Reference to the Proposed Alaska Highway Gas Pipeline.* Anchorage: University of Alaska, October 1978.

Tussing, Arlon R., and Connie C. Barlow (for the Alaska State Legislature). *The Alaska Highway Gas Pipeline: A Look at the Current Impasse.* Anchorage: University of Alaska, January 1979. (Reprinted in the *Congressional Record*, 7 February 1979, p. S1294.)

Tussing, Arlon R., and Connie C. Barlow (for the Alaska State Legislature). *Financing the Alaska Highway Gas Pipeline: What Is to Be Done?* Anchorage: University of Alaska, April 1979.

Tussing, Arlon R., and Connie C. Barlow (for the U.S. General Accounting Office). *The Struggle for an Alaska Highway Gas Pipeline: What Went Wrong?* Washington, D.C.: U.S. Government Printing Office, March 1983.

U.S. Comptroller General, General Accounting Office. *Issues Facing the Future Use of Alaskan North Slope Natural Gas.* Washington, D.C.: U.S. Government Printing Office, May 1983.

## CHAPTER 5

American Gas Association. *Gas Rate Fundamentals*, third edition. Arlington, Va.: American Gas Association, 1978.

American Gas Association, ed. *Regulation of the Gas Industry.* New York: American Gas Association, 1981.

Beard, William. *Regulation of Pipe Lines as Common Carriers.* New York: Columbia University Press, 1941.

Breyer, Stephen. *Regulation and Its Reform.* Cambridge, Mass.: Harvard University Press, 1982.

Breyer, Stephen, and Paul W. MacAvoy. "The Natural Gas Shortage and the Regulation of Natural Gas Producers." *Harvard Law Review* 86 (1973):941.

Breyer, Stephen, and Paul W. MacAvoy. *Energy Regulation by the Federal Power Commission.* Washington, D.C.: Brookings Institute, 1974.

Brown, Keith C., ed. *Regulation of the Natural Gas Producing Industry.* Baltimore, Md.: Resources for the Future and Johns Hopkins University Press, 1972.

Energy Information Administration, U.S. Department of Energy. *A Chronology of Major Oil and Gas Regulations.* Washington, D.C.: U.S. Government Printing Office, February 1982.

_____ . *Gas Supplies of Interstate Natural Gas Pipeline Companies, 1978.* Washington, D.C.: U.S. Government Printing Office, April 1980.

_____ . *Intrastate and Interstate Supply Markets under the Natural Gas Policy Act.* Washington, D.C.: U.S. Government Printing Office, October 1981.

Federal Trade Commission. *Summary Report of the Federal Trade Commission to the Senate of the United States Pursuant to S.R. 83 on Holding and Operating Companies of Electric and Gas Utilities, S. Doc. No. 73-A.*

Hooley, Richard W. *Natural Gas Financing.* New York: Columbia University Press, 1958.

Hooley, Richard W. *Financing the Natural Gas Industry.* New York: Columbia University Press, 1961.

Jacoby, Henry D., and Arthur W. Wright. *Natural Gas Price Policy: Loosening the Gordion Knot.* Cambridge, Mass.: M.I.T. Press, March 1982.

Kahn, Alfred E. *Economics of Regulation: Principles and Institutions, Volume II: Institutional Issues.* New York: John Wiley and Sons, Inc., 1971.

Kitch, Edmund W. "Regulation of the Field Market for Natural Gas by the Federal Power Commission." *Journal of Law and Economics* 11 (1968):243-80.

Leston, Alfred M., John A. Chrichton, and John C. Jacobs. *The Dynamic Natural Gas Industry.* Norman: University of Oklahoma Press, 1960.

MacAvoy, Paul W., and Robert S. Pindyck. *Price Controls and the Natural Gas Shortage.* Washington, D.C.: American Enterprise Institute, 1975.

Mitchell, Edward J., ed. American Enterprise Institute. *The Deregulation of Natural Gas.* Washington, D.C.: AEI, 1983.

Peebles, Malcolm W. *Evolution of the Gas Industry.* New York: New York University, 1980.

Pierce, Richard J., Jr. *Natural Gas Regulation Handbook.* New York: Executive Enterprises Publications Company, Inc., 1980.

Priest, A.J.G. *Principles of Public Utility Regulation*, 2 vols. Charlottesville, Va.: The Michie Company, 1969.

Sanders, M. Elizabeth. *The Regulation of Natural Gas: Policy and Politics, 1938-1978.* Philadelphia: Temple University Press, 1981.

Stotz, Louis, and Alexander Jamison. *History of the Gas Industry*. New York: Press of Stettiner Brothers, 1938.

Tussing, Arlon R., and Connie C. Barlow. "The Rise and Fall of Regulation in the Natural Gas Industry." *Public Utilities Fortnightly*, 109, no. 5 (1982):15–23.

U.S. Comptroller General, General Accounting Office. *Information on Contracts between Natural Gas Producers and Pipeline Companies*. Washington, D.C.: U.S. Government Printing Office, February 1983.

U.S. Senate, Committee on Interior and Insular Affairs, National Fuels and Energy Policy Study. *Natural Gas Policy Issues and Options: A Staff Analysis*, Serial No. 93–20 (92–55). Washington, D.C.: U.S. Government Printing Office, 1973.

Watkins, Alfred J. "Big Oil vs. Cheap Gas." *Washington Monthly*. January, 1983.

## CHAPTER 6

American Gas Association. *Gas Energy Review* (September 1982).

_____ . *Gas Facts: 1981 Data*. Arlington, Va.: American Gas Association, 1982.

_____ . *Historical Statistics of the Gas Utility Industry, 1966–1975*. Arlington, Va.: American Gas Association, 1977.

American Petroleum Institute. *Finding and Producing Oil*. Dallas, Tex.: American Petroleum Institute, 1939.

_____ . *History of Petroleum Engineering*. Dallas, Tex.: American Petroleum Institute, 1961.

Berger, Bill D., and Kenneth E. Anderson. *Modern Petroleum: A Basic Primer of the Industry*. Tulsa, Okla.: The Petroleum Publishing Company, 1978.

British Petroleum Company, Ltd. *Our Industry Petroleum*. London: British Petroleum Company, Ltd., 1970, and later editions.

Congressional Research Service (for the House Committee on Science and Astronautics). *Energy Facts*. Washington, D.C.: U.S. Government Printing Office, November 1973.

Congressional Research Service (for the Senate Committee on Foreign Affairs). *Mexico's Oil and Gas Policy: An Analysis*. Washington, D.C.: U.S. Government Printing Office, December 1978.

Energy Information Administration, U.S. Department of Energy. *Statistics of Interstate Natural Gas Pipeline Companies*. Washington, D.C.: U.S. Government Printing Office, October 1981.

Energy Information Administration, U.S. Department of Energy. *The Current State of the Natural Gas Market*. Washington, D.C.: U.S. Government Printing Office, December 1981.

Grayson, George W. *The Politics of Mexican Oil*. Pittsburgh: University of Pittsburgh Press, 1980.

Ikoku, Chi U. *Natural Gas Engineering: A Systems Approach*. Tulsa, Okla.: PennWell Publishing Company, 1980.

International Energy Agency. *Natural Gas Prospects to 2000*. France: International Energy Agency, 1982.

Interstate Oil Compact Commission. *A Study of Conservation of Oil and Gas in the United States*. Oklahoma City, Okla.: IOCC 1964.

Loftness, Robert L. *Energy Handbook*. New York: Van Nostrand Reinhold, 1978.

National Petroleum Council. *Unconventional Gas Sources*. Washington, D.C.: National Petroleum Council, December 1980.

Nehring, Richard (of the Rand Corporation, prepared for the U.S. Geological Survey). *The Discovery of Significant Oil and Gas Fields in the United States*. Washington, D.C.: U.S. Government Printing Office, January 1981.

Rand Corporation (for U.S. Department of Energy). *Mexico's Petroleum and U.S. Policy: Implications for the 1980s*. Washington, D.C.: U.S. Government Printing Office, June 1980.

Riva, Joseph P. *World Petroleum Resources and Reserves*. Boulder, Colo.: Westview Press, 1983.

Satriana, M., ed. *Unconventional Natural Gas: Resources, Potential, and Technology*. Park Ridge, N.J.: Noyes Data Corp., 1980.

Schanz, John J., and Joseph P. Riva, Jr. (for the Congressional Research Service). *Exploration for Oil and Gas in the United States: An Analysis of Trends and Opportunities*. Washington, D.C.: U.S. Government Printing Office, September 1982.

Tiratsoo, E.N. *Natural Gas*, third edition. Houston, Tex.: 1979.

Tussing, Arlon R. "Finiteness of Petroleum Resources: History and Mythology." *The Northern Engineer* 4 (1966):12–15.

Tussing, Arlon R., and Lois S. Kramer. *Hydrocarbons Processing: A Primer for Alaskans*. Anchorage: University of Alaska, August 1981.

U.S. Bureau of Mines. *Minerals Yearbook*. Washington, D.C.: U.S. Government Printing Office (annual publication).

## CHAPTER 7

American Gas Association. *Gas Facts: 1981 Data*. Arlington, Va.: American Gas Association, 1982.

Energy Information Administration, U.S. Department Of Energy. *Industrial Energy Consumption: Report on Alternative-Fuel Burning Capabilities of Large Boilers in 1979*. Washington, D.C.: U.S. Government Printing Office, February 1982.

_____ . *Natural Gas Producer/Purchaser Contracts and Their Potential Impacts on the Natural Gas Market*. Washington, D.C.: U.S. Government Printing Office, June 1982.

_____ . *An Analysis of Post-NGPA Interstate Pipeline Wellhead Purchases*. Washington, D.C.: U.S. Government Printing Office, September 1982.

_____ . *Intrastate and Interstate Supply Markets under the Natural Gas Policy Act*. Washington, D.C.: U.S. Government Printing Office, October 1981.

_____ . *The Current State of the Natural Gas Market*. Washington, D.C.: U.S. Government Printing Office, December 1981.

First Boston Research. *Large Volume Sales of Natural Gas: Their Importance and Vulnerability*. Boston: First Boston Research, August 1982.

General Accounting Office. *Natural Gas Price Increases: A Preliminary Analysis*. Washington, D.C.: U.S. Government Printing Office, December 1982.

Hilt, Luis. "Chronology of the Gas Industry." *American Gas Journal* (May 1950).

Stotz, Louis, and Alexander Jamison. *History of the Gas Industry*. New York: Press of Stettiner Brothers, 1938.

Tussing, Arlon R. "An OPEC Obituary." *The Public Interest* 70 (1983): 3–21.

Tussing, Arlon R., and Connie C. Barlow (for the U.S. Department of Energy). *Supplemental Gas Marketing and Financing Issues*. Washington, D.C.: U.S. Government Printing Office, October 1978.

_____ . (for the Alaska State Legislature). *An Introduction to the Gas Industry with Special Reference to the Proposed Alaska Highway Gas Pipeline*. Anchorage: University of Alaska, October 1978.

_____ . "A Survival Strategy for Gas Companies in the Post-OPEC Era." *Public Utilities Fortnightly* 3 (1983):13–18.

U.S. Comptroller General, General Accounting Office. *Information on Contracts between Natural Gas Producers and Pipeline Companies*. Washington, D.C.: U.S. Government Printing Office, February 1983.

## CHAPTER 8

American Gas Association. *Gas Energy Review* (May 1981).

Bonbright, James C., and Gardiner C. Means. *The Holding Company*. New York: McGraw-Hill Book Company, Inc., 1932. Reprint. New York: Augustus M. Kelly Publishers, 1969.

Energy Information Administration, U.S. Department of Energy. *An Analysis of Post-NGPA Interstate Pipeline Wellhead Purchases*. Washington, D.C.: U.S. Government Printing Office, September 1982.

Hooley, Richard W. *Natural Gas Financing*. New York: Columbia University Press, 1958.

_____ . *Financing the Natural Gas Industry*. New York: Columbia University Press, 1961.

Leston, Alfred M., John A. Chrichton, and John C. Jacobs. *The Dynamic Natural Gas Industry*. Norman: University of Oklahoma Press, 1963.

Moody's Investors Service. *Moody's Public Utility Manual, 1982*. New York: Moody Investors Service, 1982.

Nevner, Edward J. *The Natural Gas Industry: Monopoly and Competition in Field Markets*. Norman: University of Oklahoma Press, 1960.

Peebles, Malcolm W.H. *Evolution of the Gas Industry*. New York: New York University, 1980.

Priest, A.J.G. *Principles of Public Utility Regulation*. 2 vols. Charlottesville, Va.: The Michie Company, 1969.

Ritchie, Robert F. *Integration of Public Utility Holding Companies*. Ann Arbor: University of Michigan, 1955.

# GLOSSARY

**abandonment**—in utility law, governmental authorization for a utility to cease provision of a particular service and/or to shut down a particular facility that had been part of a company's "rate base."

**abiotic (abiogenic) theory**—a theory of hydrocarbon genesis that attributes the origins of a substantial portion of the earth's endowment of natural gas to inorganic processes; that is, methane is believed to have been one of the earth's primeval gases, which predated the beginnings of life and which remained trapped (and unoxidized) within the earth as the crust cooled. COMPARE: **biotic theory**.

**advance payments**—a program authorized by the FPC during the 1970s gas shortages by which interstate pipelines channeled interest-free loans (financed by ratepayers) to gas exploration companies contingent upon dedication of discoveries to that particular pipeline or full reimbursement in the event of dry holes.

**affiliated companies**—companies, one of which owns a controlling or influential share of the other(s) or all of which are controlled by the same parent company. Federal regulators historically have subjected wellhead prices of gas producers to greater oversight if the producer is affiliated to the purchasing gas pipeline.

**AFUDC**—allowance for funds used during construction of utility facilities. The utility accrues a return on its invested capital during the construction period, which then is added to the direct cost of the facility for the purpose of determining the **rate base** on which it is allowed to earn a return after the facility goes into operation. While collection of AFUDC from consumers is thus deferred, the equity component is treated as a current profit on the company's books. COMPARE: **CWIP** (construction work in progress).

**allowables**—standard production rates assigned to individual oil and gas wells by a state oil and gas conservation agency for the purpose of **market-demand prorationing.** When producing capacity exceeds demand, the agency limits production from each well to a specified allowable percentage of officially recognized capacity.

**anticline**—a geologic feature in which the constituent rock strata have been compressed laterally into an inverted U-shaped fold. Anticlines are frequently **traps** for oil and gas. Structural counterparts of anticlines are "synclines," which are concave folds, as viewed from above.

**areawide pricing**—a method the FPC employed during the 1960s to establish wellhead-price ceilings for interstate gas producers on a regional (rather than a case-by-case or a national) basis.

**arms-length transactions**—transactions between unaffiliated companies; for example, sales by a gas producer to a separate pipeline company.

**associated gas**—natural gas found in association with crude oil, either as "dissolved" or "solution" gas within the oil-bearing strata or as "gas cap" gas just above the oil zone. COMPARE: **nonassociated gas.**

**balance-sheet financing**—a conventional method for financing capital projects whereby a company solicits debt on the basis of its overall financial strength and commits its companywide assets as security. COMPARE: **nonrecourse project financing.**

**base-load**—The lowest load level during a facility's daily or annual cycle; supplies of natural gas intended to service customer needs on a year-round basis. COMPARE: **peak-shaving.**

**bcf**—billion cubic feet.

**biomass**—any body or accumulation of organic material. In the gas industry, biomass refers to the organic waste products of agricultural processing, feedlots, timber operations, or urban refuse from which methane can be derived.

**biotic (biogenic) theory**—a theory of hydrocarbon genesis that attributes the origins of the earth's endowment of natural gas (and other hydrocarbons) exclusively to biological processes; that is, methane is believed to be a product of organic decomposition of the tissues of once-living plants and animals. COMPARE: **abiotic theory.**

**boiler fuel**—fuels suitable for the generation of steam (or hot water) in large industrial or electric-utility boilers. Natural gas, residual oil, coal, and uranium are the dominant boiler fuels.

**border price**—the official price for gas sold at the U.S./Canadian border, as determined by the Canadian government in consultation (in theory) with the United States.

**British Thermal Unit (btu)**—the standard measurement for heat employed in the U.S. gas industry. One btu raises the temperature of one pound of water by one degree Fahrenheit from 58.5 to 59.5 degrees under standard pressure of 30 inches of mercury. Natural gas of "pipeline quality" contains about 1,000 btu per cubic foot.

**burner-tip**—signifying delivery to the final customer. A burner-tip price, for example, is the price charged the end user.

**butane**—a hydrocarbon component ($C_4H_{10}$) of produced natural gas and crude oil; one of the **natural-gas liquids** (NGL), and a component of liquefied petroleum gas (LPG).

**byproduct coke gas**—a fuel-rich vapor that is a byproduct of the coking process. Also called **coke-oven gas.**

**candlepower**—an obsolete unit of luminous intensity used during the early decades of the gas industry. A "candle" was originally defined in terms of a wax candle with standard composition and equal to 1.02 "candelas."

**cap**—1. nonpermeable rock found above oil and gas reservoirs, which through geologic time prevented these hydrocarbons from dissipating. 2. SEE: **gas cap.**

**capital structure**—pertaining to a company's invested capital; specifically, the relative proportions of various categories of debt and equity in the company's total capital.

**carbon black**—a substantially pure form of finely divided carbon, usually produced from liquid or gaseous hydrocarbons by controlled combustion with restricted air supply. Beginning as "lamp-black" (named for its genesis in household kerosene lamps), it was used as a pigment for printing inks and paints. Later, it became a valuable strengthening agent for rubber products.

**carburet(t)ed water gas**—an advanced technique developed in the 1870s by Thaddeus Lowe for the production of coal gas.

**casinghead gas**—natural gas that flows from an oil well along with the liquid petroleum. It is also called "associated," "dissolved," or "solution" gas because it resides beneath the earth's surface in conjunction with crude oil. Casinghead gas is distinguished from "gas cap gas," which is another form of "associated" gas that accumulates above the oil-bearing strata and which is therefore produced from wells that do not tap the oil resource.

**certificate**—"certificate of public convenience and necessity"; a permit or license granted to a public utility by a regulatory commission, authorizing the utility to provide a specific service or build and operate a specific facility. SEE: **franchise.**

**charter**—a document issued by a sovereign, legislature, or other authority, creating a public or private corporation. In the nineteenth century, before state legislatures passed statutes allowing blanket rights of incorporation, prospective businesses had to obtain charters on a case-by-case basis.

**city gate**—a location at which gas ownership passes from a gas pipeline company to a local distributor.

**coal gas**—gas manufactured from coal.

**coke**—the solid carbonaceous residue produced from the destructive distillation of coal or oil; a vital material for steelmaking.

**coke-oven gas**—a fuel-rich vapor that is a byproduct of the coking process. Also called **byproduct coke gas.**

**commodity charge (energy charge)**—a customer charge for utility service that is proportional to the amount of gas (electricity, etc.) actually purchased. COMPARE: **demand charge.**

**common carrier**—a transporter obligated by law to provide service to all interested parties without discrimination to the limit of its capacity. If the capacity of a common-carrier pipeline is insufficient to fulfill demand, it must offer services "ratably" to all shippers in proportion to the amounts they tender for shipment. In the United States, most oil pipelines are required to operate as common carriers, but gas pipelines are primarily **private carriers** and secondarily **contract carriers.**

**common law**—an accepted legal stricture of general and universal application, based on customs and traditional usages (beginning in England) rather than upon codified written laws.

**common purchaser**—an oil or gas carrier that is required by law to purchase without discrimination from all parties tendering oil or gas produced from a given reservoir, field, or area. If the amount of oil or gas tendered to a common-purchaser pipeline exceeds the quantity it wants to buy, the pipeline must allocate its purchases among the various producers **pro rata** (ratably) according to the amounts tendered.

**compressed natural gas (CNG)**—natural gas that is highly compressed (though not to the point of liquefaction), so that it can be utilized by an operation not attached to a fixed pipeline. CNG is already used extensively as a transportation fuel for automobiles, trucks, and buses in some parts of Italy, in New Zealand, and in Western Canada, and has recently begun to penetrate some regions of the United States.

**constant dollars**—with respect to past or future costs or prices, figures that are quoted in inflation-adjusted terms relative to a particular year. For example, cumulative construction costs incurred between 1975 and 1980 can be referenced in "1980 dollars." "Real prices" are measured in constant dollars. COMPARE: **current dollars.**

**consumer guarantee**—forms of interstate gas-pipeline tariffs developed during the 1970s that committed downstream purchasers to make minimum payments in order to assure recovery of the capital invested in supplemental gas projects financed on a "nonrecourse" basis.

**contract carrier**—a transporter (such as a gas pipeline company) that voluntarily provides its service on a contractual basis for other parties. COMPARE: **common carrier, private carrier.**

**conventional financing**—standard and accepted methods for financing capital projects. Includes both balance-sheet and project financing, but does not apply to nonrecourse forms of project financing.

**conventional gas**—gas that can be produced under current technologies at a cost that is no higher than its current market value. COMPARE: **unconventional gas.**

**cost of capital**—the threshold interest rate and return on equity that a particular utility must offer prospective bondholders and stockholders in order to attract debt and equity investors.

**cost-of-service**—the fundamental principle of gas utility regulation, whereby customer charges are based upon the actual costs of providing the service (including a competitive profit), rather than allowing prices to rise to whatever customers may be willing to pay. COMPARE: **value-of-service pricing.**

**cost-of-service tariff**—a form of gas-transmission tariff proposed during the 1970s to support financing of supplemental gas projects. Unlike the standard **fixed-price tariff,** a utility operating under a cost-of-service tariff is allowed to adjust its customer rates on a monthly basis to track the actual costs of providing that service—including per-unit adjustments attributable to a loss of load.

**correlative rights**—multiple ownership of oil and/or gas within a common reservoir. State conservation agencies are empowered to protect correlative rights, using such tools as well-spacing rules, and in most states, compulsory-**unitization** and **common-purchaser** laws.

**covenants**—clauses inserted in debt instruments (such as bonds) that place restrictions on the company receiving funds so that the purchaser of the bond (the lender) has greater security against a financial loss.

**cryogenic**—pertaining to a superinsulated vessel for storing liquefied natural gas or for transporting it by truck or tanker.

**current dollars** (or **nominal dollars**)—with respect to past or future costs and prices, figures that are not adjusted for inflation. Data for each year are stated at the dollar value then (or expected to be) current. COMPARE: **constant dollars.**

**curtailment policies**—state and federal regulatory policies that provide guidance (and legal protection) to gas utilities that must adopt priorities and design schedules for curtailing service to firm customers during a supply shortfall.

**CWIP**—construction work in progress. Utility property that, though still under construction, is allowed to be placed in the rate base. Most regulatory commissions do not permit CWIP but, rather, require the utility to accrue **AFUDC** (allowance for funds used during construction) and add it to the rate base only when the facility goes into operation.

**death spiral**—a self-propagating collapse in utility revenues, triggered by a company's attempt to respond to a price-induced loss in sales by charging higher prices (in an attempt to maintain revenues), thereby prompting further load loss.

**debt/equity ratio**—the relative proportion of debt and equity comprising the "capital structure" of a company.

**declining block rates**—a utility rate structure by which customers who consume greater quantities of gas are charged lower per-unit rates, which descend in a step-like fashion. COMPARE: **inverted rate structure.**

**deep gas**—natural gas located 15,000 feet or more below the earth's surface, as defined (and deregulated) in Section 107 of the Natural Gas Policy Act of 1978.

**deficiency payments**—payments by interstate gas pipeline companies to gas producers in fulfillment of "take-or-pay" contract terms. An obligation to make a deficiency payment is triggered when a pipeline is unable (or unwilling) to take a specified minimum volume of gas set forth in a supply contract, and it is therefore obligated to pay for the deficient volume of gas.

**deliverability**—the amount of gas that a pipeline or producer is able to deliver, limited either by the terms of its supply contracts or its own plant capacity.

**demand charge**—a customer charge for utility service that reflects the extent to which a particular customer chooses to purchase a right to draw a certain maximum (or unlimited) volume of gas at any time during the year. Customers who purchase gas on an "interruptible" basis do not, therefore, pay a demand charge. COMPARE: **commodity charge.**

**demonstration plants**—commercial-scale synthetic gas plants proposed in the 1970s to manufacture high-btu coal gas on a year-round basis.

**depreciation**—in accounting terms, the loss in value of a particular piece of physical property through time. Utilities are allowed to recover (or "amortize") their investment in physical plant in accordance with an approved depreciation schedule.

**depreciation schedule**—the mathematical basis for calculating the rate at which a utility will recover its investment in physical plant. "Straight-line" depreciation by which investment is recovered in equal increments during the period established for full-cost recovery is the most common form accepted by state and federal regulators. Four percent straight-line depreciation, for example, means that the investment is recovered in equal increments over a 25-year period.

**Devonian shales**—shales originating from the muds of shallow seas deposited about 350 million years ago, during the Devonian period of the Paleozoic era. Devonian shales from which "unconventional gas" can be drawn are found relatively close to the earth's surface in parts of the Great Lakes region and areas to the west of the Appalachian divide.

**direct sales**—transactions in which a gas pipeline sells its gas directly to an end-user, like a large industrial plant, rather than to a distributor for resale. Direct sales by interstate gas pipelines are not subject to federal rate regulation under the Natural Gas Act.

**dissolved gas**—a form of associated gas found in solution with petroleum and, therefore, produced from oil wells as **casinghead gas.**

**domicile**—legal residence; the state in which a company has secured its corporate status.

**dual-fuel capacity**—the ability of an energy consumer (foremost, large industrial and electric-utility customers) to utilize two kinds of fuels. Multi-fuel capacity allows even greater flexibility.

**economic cost**—SEE: **resource cost.**

**economic rent**—the difference between the market value of a particular commodity and its **resource cost** (or economic cost).

**emergency purchases**—a federal program operative during the gas shortages of the 1970s whereby the FPC authorized an industrial customer or distributor to purchase gas directly in the field on a shorterm basis at prices above regulated ceilings and to arrange transportation for such gas with an interstate pipeline on a "contract" basis.

**eminent domain**—the right of a government to appropriate private property for public use, usually with compensation to the owner.

**ethane**—a hydrocarbon component ($C_2H_6$) of produced natural gas and to a lesser extent of pipeline gas. One of the **natural-gas liquids (NGL).**

**exempt gas**—categories of gas exempted from federal price ceilings, pursuant to section 107 of the Natural Gas Policy Act of 1978, because physical characteristics of the gas implied high production costs. Most of the exempt gas that was produced following passage of the NGPA was "deep" gas, found below 15,000 feet.

**FERC**—the Federal Energy Regulatory Commission, the federal agency that regulates interstate gas pipelines and interstate gas sales under the Natural Gas Act. Successor to the Federal Power Commission **(FPC)**, FERC is considered an independent regulatory agency responsible primarily to Congress, but it is housed in the Department of Energy.

**firm service**—gas that is sold with a guarantee for delivery. Customers generally pay more for firm gas than for "interruptible" gas, which the utility may curtail at its discretion without liability.

**fixed costs (embedded costs)**—that portion of the total cost of any business activity that cannot be reduced by reducing the level of service. Fixed costs are primarily depreciation of the "sunk" capital invested in physical plant.

**fixed-price tariff** (or **stated-rate tariff**)—the standard form of gas-utility tariff whereby customer charges are calculated for the "test year" in question. With the exception of **purchased-gas-adjustment** (PGA) and fuel-adjustment clauses, customer charges remain fixed until the company (or some **intervenor** such as the public utility commission staff) successfully petitions its regulator for any other adjustment to reflect changed circumstances. COMPARE: **cost-of-service tariff.**

**fixed-variable rate structure**—a rate structure for a utility tariff that allocates all of the "fixed costs" to the "demand" component of customer chargers. COMPARE: **United formula; Seaboard formula.**

**FOB**—"free on board." In the international LNG trade, an FOB price applies to the delivered product at the terminal of the exporting nation, while a CIF ("cost, insurance, and freight") price attaches to the product when delivered at the importing terminal.

**force majeure**—a legal principle by which contractual obligations are waived if a "superior force" (such as the weather, a war, or "an act of God") makes it impossible for those obligations to be met.

**forward contract**—a contract for future delivery at a price determined in advance.

**FPC**—the Federal Power Commission, the agency that regulated interstate gas pipelines and interstate gas sales from 1938 to 1977. Succeeded by **FERC** (the Federal Energy Regulatory Commission).

**franchise**—1. (noun) a special privilege conferred by a government on an individual or a corporation to utilize public ways and streets. 2. (verb) the act of issuing a franchise. SEE: **certificate.**

**fuel cell**—a manufactured device that converts natural gas and other gaseous fuels into electricity and heat via an electrochemical process that avoids the inefficiencies of ordinary combustion.

**gas cap**—gas-rich strata overlying the oil-bearing strata of a petroleum reservoir. A gas cap forms when the ratio of gas to oil in a particular reservoir is too high for all of the gas to remain in solution with the oil. Gas-cap gas is produced from wells distinct from those producing oil and casinghead gas.

**gas liquids** (also, **natural-gas liquids** or **NGLs**)—the hydrocarbon components of "wet gas" whose molecular structures are heavier than methane but lighter than crude oil. Gas liquids include ethane ($C_2H_6$), propane ($C_3H_8$), and butane ($C_4H_{10}$).

**gas works**—the facilities in which manufactured gas was produced in the earliest decades of the gas industry before natural gas became widely available.

**gathering systems**—pipelines owned and operated by gas producers that are considered an integral part of gas production (rather than transmission) and are therefore usually exempt from state and federal utility regulation.

**geographic integration**—establishment by a single company (or affiliated companies) of a business presence over a wide geographic area.

**geopressurized brines**—saltwater found in underground rock strata for which reservoir pressures are far higher than commonly exists at such depths. In the United States, methane-rich geopressurized brines are found at great depths in the Gulf Coast states and under the Gulf.

**heating value**—the amount of heat produced by the complete combustion of a unit quantity of fuel. Technically, heating value is expressed as either "gross" or "net" depending primarily upon whether adjustment is made for the "latent heat of vaporization" of the combustion products.

**high-priority customers**—those customers denoted in utility curtailment schedules whose service is the last to be affected by supply or capacity shortfalls. Residential and small commercial customers, along with schools and hospitals, are high priority customers under virtually all pipeline and distributor curtailment plans.

**Hinshaw pipeline**—a pipeline subject to state rather than federal jurisdiction under the Hinshaw Amendment to the Natural Gas Act because, while it sells gas in "interstate commerce," its physical plant lies entirely within the consuming state.

**holding company**—a company that holds a controlling ownership interest in two or more companies. Holding companies that control one or more public utility companies engaged in interstate commerce are regulated by the Public Utility Holding Company Act of 1935.

**horizontal integration**—within the energy business, involvement by a single company (or affiliated companies) in the production, processing, or transportation of more than one form of energy—as in gas and electricity.

**hydrates**—SEE: **methane hydrates.**

**incandescent**—a method of illumination that utilizes the heat-induced glow of a solid (like the filament of an electric lamp) rather than the light of a flame.

**incentive rate of return (IROR)**—a regulatory concept prescribed for the Alaska Highway gas pipeline by which the sponsors were to be rewarded or penalized according to their success in keeping construction costs within the projected budget.

**incremental pricing**—1. generally, a form of utility pricing whereby the cost of a particular gas acquisition (usually high-cost gas) is allocated to specific customers or classes of service, rather than "rolled-in" with other gas purchases. Incremental pricing was considered (though rejected) by federal regulators during certification hearings for a variety of supplemental-gas projects during the 1970s. 2. In the Natural Gas Policy Act of 1978, incremental pricing refers to the allocation to certain classes of industrial customers (within an overall regime of rolled-in pricing) of **increases** in gas costs resulting from certain kinds of gas purchases.

**independent producer**—1. in gas-industry parlance, a gas producer not affiliated with its purchasing gas-pipeline company. Federal regulators have historically subjected wellhead prices charged by independent producers to less oversight than those charged by affiliated producers. 2. in oil-industry parlance, a producer not affiliated with one of the "majors" (like Exxon or Mobil). The "major" oil companies are the dominant "independent gas producers."

**interruptible sales**—gas that is sold without a gurantee of delivery. Gas utilities curtail their interruptible customers in order to adjust to seasonal (or long-term) shortfalls in supply or plant capacity. Because utilities retain the right to curtail interruptible customers, gas sold under interruptible contracts is generally less expensive than "firm" sales.

**intervenor**—a person, business entity, or public body that is granted the right to participate in hearings or examinations on matters primarily affecting other parties.

**intrastate**—within the boundaries of a single state.

**inverted rate structure (graduated rate)**—a rate schedule that assigns increasing per-unit charges for higher levels of per-customer gas consumption. Inverted rate structures have become increasingly fashionable since the 1970s gas shortages as a device to promote conservation. COMPARE: **declining block rate.**

**joint costs**—those costs for gas and oil production that are inextricably linked by the fact that **associated gas** is produced in conjunction with oil.

**joint venture**—a business enterprise undertaken by two or more firms acting as equity partners. In the gas industry, joint ventures usually involve high-cost pipeline or supplemental-gas project construction beyond the financial capability of a single firm.

**kerosene**—a "middle distillate," heavier than naphtha, drawn from crude oil. It was an important illuminant before electric lighting became available, and today it is used for diesel fuels, home heating oil, and certain grades of jet fuels.

**leverage**—a financial term (used either as a noun or a verb) that signifies a company's attempt to bolster its equity capital with an infusion of debt capital.

**lien**—the right to take or hold the property of a debtor as security or payment for a debt.

**life-cycle costing**—a technique for evaluating the merits of a particular investment by which consideration is given to the costs (and the benefits) of that investment over its entire serviceable life.

**lifeline rates**—a ratemaking arrangement under which a minimum amount of gas or electricity deemed essential to life is provided to residential customers at rates below average system costs. Lifeline rates are subsidized by higher rates imposed on residential consumption above the minimum or on other customer classes.

**life-of-the-field contract**—a commitment by a producer to sell gas to a buyer for as long as the "dedicated" field is capable of production.

**liquefied natural gas (LNG)**—natural gas that is chilled below its boiling point ($-258.7°$Fahrenheit, or $-161.5°$ Celsius) so that it can be stored in liquid form, thereby occupying 1/625 of the space that it requires at ambient temperatures and pressures.

**load factor**—the ratio of average- to peak-day use calculated over the course of an entire year. It is considered wise utility management for a company to strive for a high load factor (as close to "1" as possible). COMPARE: **plant factor.**

**load management**—techniques employed by utilities to reduce wide seasonal variations in customer demand and to thereby improve "load factors."

**load upgrading**—a strategy by which gas utilities increase the proportion of their total sales to residential and small-commercial customers (who are willing to pay higher prices) at the expense of industrial sales, particularly to interruptible customers. Load upgrading usually involves investment in gas-storage capacity (to deal with the greater seasonal variation in residential-commercial demand) and extension of gas mains into lower density neighborhoods.

**looping**—increasing the capacity of a transmission system by laying an additional pipe beside the original.

**low-priority users**—those customers denoted in utility curtailment schedules whose service is the first to be affected by a supply shortfall. Generally, large industrial plants and electric utilities with alternate-fuel capability are classed as low priority.

**manufactured gas**—energy-rich vapors produced from controlled thermal decomposition or distillation of hydrocarbon feedstocks, including coal, oil, and coke-oven feedstocks. Manufactured gas historically was a low-btu grade (less than 500 btu per cubic foot). Today, however, only high-btu grades (sometimes called synthetic natural gas or SNG) are suitable for transport in the nation's pipeline system.

**marginal users**—large industrial and electric-utility customers for which fuel-switching between gas and residual oil or coal poses little difficulty. Marginal users are the most price-sensitive gas consumers.

**market-clearing price**—the price at which supply and demand are in balance with respect to a particular commodity at a particular time. A market-clearing price is high enough to prevent a shortage but low enough to ensure that all supplies then available can be sold.

**market-demand prorationing**—state regulation of oil or gas producers intended to prevent surplus production, by allocating current demand *ratably* among all wells in a reservoir, field or statewide. SEE: **prorationing.**

**market-out clause**—a contractural term that gives the buyer a right to terminate the agreement if unable to market the gas at the contractually determined price. During the 1970s gas shortages, few new transactions included market-out provisions; when a supply surplus developed in 1982, however, the insertion of market-out clauses became common practice.

**mcf**—thousand cubic feet. One mcf has a heating value of approximately one million btu **(mmbtu).**

**methanation**—1. a relatively simple form of controlled organic decomposition (primarily of urban sludge and solid waste) that yields methane vapors. 2. upgrading of low-btu gas manufactured from coal (whose components are mainly hydrogen and carbon monoxide) to a higher-btu gas composed mainly of methane.

**methane**—$CH_4$, the primary component of natural gas; the lightest of all hydrocarbons.

**methane hydrates**—gas frozen into a physical-chemical bond with water. Methane hydrates occur naturally in permafrost regions of the Arctic and deep sea sediments.

**methanol**—$CH_3OH$, the simplest of the alcohols; usually manufactured from methane, though it can be made from other hydrocarbons through more complex processes ("wood alcohol").

**minimum-bill tariff**—a utility tariff that commits the downstream purchaser (usually a distribution company) to make payments sufficient to meet debt service obligations (and sometimes a share of equity recovery) in the event that a large capital project temporarily or permanently ceases to function.

**mmbtu**—million btu.

**monopsony**—a market condition in which a single buyer faces a large number of sellers; obverse of monopoly. In the early days of the natural-gas industry, individual pipeline companies usually held monopsony positions in their gas-supply areas.

**naphtha**—one of the lighter constituents of crude oil, from which gasoline products ("light distillates") are drawn. Naphtha was the dominant feedstock for oil-gas forms of manufactured gas and it is still a preferred feedstock for **peak-shaving** plants that produce **substitute natural gas (SNG).**

**natural gas**—a naturally occurring mixture of energy-rich vapors (primarily methane) found beneath the earth's surface, often in association with petroleum. Natural gas of "pipeline quality" contains about 1,000 btu per cubic foot.

**natural gas liquids (NGLs)**—SEE: **gas liquids.**

**natural monopoly**—an activity (such as the provision of gas, water, and electrical service) characterized by "economies of scale" wherein cost of service is minimized if a single enterprise is involved. Because duplicative facilities are economically inefficient, enterprises that are natural monopolies tend to attract business organizations that are in fact monopolistic; competition is retained only by governmental mandate or, conversely, the monopoly is sanctioned but regulated through issuance of a franchise.

**nonassociated gas**—natural gas that is found in a reservoir free of crude oil.

**noncompletion risk**—the risk that a project will be abandoned after substantial capital has been invested. The noncompletion risk attendant to billion-dollar supplemental-gas projects was an obstacle to project financing.

**nonjurisdictional customers**—large electric utilities and industrial plants that purchase gas by "direct sale" from interstate gas pipelines, rather than from local distributors. Because the transaction is not a "sale for resale," it is not subject to federal regulation of rates.

**nonrecourse project financing**—a method for financing capital projects whereby a company solicits debt for a particular project and pledges the prospective revenues as debt coverage. Because the company does not also pledge its general assets as security, nonrecourse forms of financing depend upon novel regulatory tools (such as minimum bill tariffs). A few base-load LNG projects were financed on a nonrecourse basis during the 1970s before ensuing difficulties made lenders wary of this approach.

**nonrenewable energy resource**—an energy resource that is depletable. Oil and gas are nonrenewable energy resources unless considered in the context of geologic time (wherein the products of hydrocarbon combustion will recycle into new organic tissues, lithify into rocks, and perhaps migrate as oil or gas into geologic traps once again). Solar, wind, and hydropower are, on the other hand, renewable energy resources.

**off-system sales**—sales by a gas utility to a customer outside of its currently authorized market. During the gas glut of the early 1980s, interstate pipelines sold gas off-system to other interstate pipelines as a means of shaving their own supply surpluses.

**oil gas**—manufactured gas based on naphtha feedstocks.

**old gas**—a regulatory category for natural gas delineated by federal law or administrative order. In the Natural Gas Policy Act, old gas wells are those for which deliveries commenced in 1977 or earlier.

**old-gas subsidy cushion**—the difference between the average cost of natural gas to a regulated company and the market value of that gas. During the 1970s, the average price of gas on contract to interstate pipelines was below its market value because of a combination of wellhead price controls and long-term, fixed-price contracts. The old-gas subsidy cushion gave pipelines considerable leeway to purchase supplemental or deregulated gas at prices above market value.

**original cost (historic cost)**—a method of pipeline "valuation" by which a utility's rate base (upon which a profit is allowed) is the amount of capital actually invested, less cumulative depreciation. Federal regulation of interstate gas pipelines is based on original cost, rather than "replacement" or "reproduction" cost.

**overrun pool**—A financing strategy conceived by the sponsors of the Alaska Highway gas pipeline by which an excess of capital commit ments would be sought prior to construction to ensure that capital would be available for project completion in the event of cost over-runs.

**overthrust belts**—regions in which a large block of the earth's crust has been pushed and squeezed up and over the surrounding rock. Overthrust belts are usually found in proximity to mountains, as in the Rocky Mountain and Appalachian overthrust belts.

**PSC**—a state public service commission.

**PUC**—a state public utility commission.

**peak day**—on an annual basis, the day of highest customer demand.

**peak-shaving**—methods for accommodating the seasonal periods of greatest customer demand, including the drawdown of gas held in underground storage and the production of substitute natural gas (SNG).

**Pemex**—Petroleos Mexicanos, the state petroleum company of Mexico.

**plant factor**—the extent to which a particular facility is utilized. For a gas pipeline, plant factor is the ratio of average daily throughput to design capacity. COMPARE: **load factor.**

**posted price**—in U.S. oil-industry parlance, a price that a refiner offers to pay producers in a particular field. The refiner has the flexibility to adjust its posted price to track changes in market outlook.

**prebilling clause**—a special form of tariff authorized by Congress for the Alaska Highway gas pipeline by which ratepayers would make payments to cover fixed-plant depreciation for completed segments of the system even if construction still in progress for other segments prevented shipment of Alaska gas.

**prebuild**—The first phase of construction for the Alaska Highway gas pipeline, which included southerly portions of the Canadian seg-ments and the Eastern Leg (the Northern Border pipeline) and the Western Leg in the United States.

**price-elasticity of demand**—a measure of the sensitivity of demand to changes in price: specifically, the ratio between the percentage change in sales volume and the percentage change (in the opposite direction) of the price change which causes the change in sales.

**primary energy**—energy that is not derived from any other energy source. For example, fossil-fuel generated electricity is not a primary energy source because it is generated from gas, oil, or coal.

**private carrier**—a transporter that owns the commodity it carries. In-terstate gas pipelines operate primarily as private carriers, although contract carriage business is increasing. COMPARE: **common carrier; contract carrier.**

**process-gas consumers**—those industries (like grain drying and petrochemical manufacturing) that cannot readily substitute other fuels or hydrocarbons to yield the flame characteristics or feedstock chemistry of natural gas.

**pro-forma tariff**—a tariff filed by a gas utility seeking certification of a particular project. The pro-forma tariff enables regulators to assess the cost impacts of the project, but it is generally nonbinding; upon project completion, the company is free to request approval of a tariff that reflects actual expenditures.

**project financing**—a method for financing capital projects whereby a company solicits debt for a particular project, dedicates the prospective flow of project revenues as debt coverage, and only secondarily (if at all) pledges its general corporate assets as security on the loans. SEE: **nonrecourse project financing.**

**propane**—a hydrocarbon component ($C_3H_8$) of produced natural gas; one of the **natural-gas liquids (NGLs)**. Also an oil-refinery byproduct, and the principal constituent of **liquefied petroleum gas (LPG)**.

**propane-air**—a form of **substitute natural gas (SNG)** used for peak-shaving. Propane (2,300 btu per cf) is more than twice as rich in energy as is methane, so it is diluted with air to achieve "pipeline quality" gas of thermal parity with methane.

**pro-rata** or **ratably**—with respect to a production surplus (or a surplus of shippers seeking the services of a particular common carrier pipeline), a rule by which cutbacks are made proportionally without consideration of any differences that may exist in price or other terms of the transactions.

**prorationing**—1. with respect to oil or gas production, state regulation that apportions allowable production **ratably** among wells, either to protect the correlative rights of owners tapping a single source or to prevent discrimination in the event that market demand is less than production capacity. 2. with respect to **common-carrier pipelines,** the apportionment of limited transportation capacity **ratably** among parties tendering oil or gas for shipment. SEE: **market-demand prorationing.**

**psi**—pounds per square inch.

**public utility**—a publicly or privately owned business that is subject to government regulation because it is "a business affected with a public interest" and provides an essential service (such as gas transportation and distribution) to the consuming public.

**purchased gas adjustment (PGA) clause**—a clause within a fixed-price tariff that allows a utility to automatically adjust its customer rates to reflect an increase (or a decrease) in gas purchase costs without first petitioning the regulatory body for approval. At the federal level, PGAs allow for automatic tariff adjustments twice a year, while some states allow PGA adjustments monthly.

**quad**—abbreviation for quadrillion btu. For natural gas, roughly one trillion cubic feet **(tcf).**

**ratable**—SEE: **pro-rata.**

**rate base**—the value of utility property recognized by a regulatory authority upon which the company is permitted to earn its rate of return. Generally, this value represents the "depreciated original cost" of property "used and useful" in public service.

**rate regulation**—in the gas industry, regulation of transmission and distribution utilities to ensure "just and reasonable" service charges based on "cost-of-service" principles.

**rate schedule**—a component of pipeline and distributor tariffs that sets forth the prices to be charged customer classes for the types of service available.

**regulatory gap**—1. an apparent need for regulation. 2. a phenomenon by which the implementation of a particular law or rule seems to in itself generate new problems for which additional regulation becomes necessary.

**reinject**—the process by which casinghead gas (a co-product of oil production) is returned to the petroleum-bearing strata for pressure maintenance and enhanced oil recovery.

**replacement cost**—a possible method of pipeline "valuation" by which a utility's rate base (upon which a profit is allowed) is considered to be the amount of money that would be needed to replace its physical plant with current techologies (scaled to provide the same level of service) and at current prices.

**reproduction cost**—a possible method of pipeline "valuation" by which a utility's rate base (upon which a profit is earned) is considered to be the amount of money that would now be required to replace the physical plant with no accommodation made for changes in design which might result from improved technology.

**reserves**—with respect to oil or gas, that proportion of the resource that is commercially recoverable under current economic conditions with current technology. "Proved" or "established" reserves are the portion of the resource that is in known reservoirs and that is believed to be recoverable with the highest degree of confidence.

"Indicated," "inferred," or "probable" reserves are the additional resources associated with known reservoirs that are expected (in the statistical sense) to be recoverable. "Speculative" or "possible" reserves are those resources (in addition to the foregoing), outside the vicinity of known reservoirs, which are expected to be recoverable. "Ultimate reserves" or "ultimate recoverable resources" are the sum of proved, indicated, and speculative reserves.

**reserves-to-production (R/P) ratio** (or **life index**)—for a particular gas field or fields, the ratio of remaining recoverable reserves to the current annual rate of production. For a gas pipeline or pipelines, the R/P ratio is the ratio of dedicated gas reserves to the current rate of annual sales. An R/P ratio of 20, therefore, means that sufficient gas remains for a field to continue producing or for a pipeline to continue delivering gas for twenty years.

**reservoir**—a rock formation or trap holding an accumulation of crude oil and/or natural gas.

**residual oil**—the heavier hydrocarbons contained in crude oil, which do not boil off in the distillation process. Because of its viscosity, heterogenity, impurities, and concentrations of sulphur, these "bottoms" (sometimes called No. 6 oil) are burned primarily in electric-utility and industrial boilers or used as bunker fuel on the high seas.

**resource cost (economic cost)**—the cost of capital, materials, and labor necessary to produce and market a particular resource. For an enterprise to be profitable, resource costs must be no higher than market value.

**rolled-in pricing**—a form of cost accounting whereby a utility charges its customers the weighted average cost of all gas acquisitions, rather than allocating specific purchases to specific customer classes or geographic markets. Rolled-in pricing is standard gas-industry practice. COMPARE: **incremental pricing.**

**Rule of Capture**—a method of defining resource ownership by which ownership is contingent upon physical possession. In the case of natural gas and oil, the Rule of Capture was a powerful incentive for wasteful production techniques—including excessively dense drilling in the rush to produce the resource before one's neighbor claimed it. Conservation rules imposed by state governments and voluntary and mandatory **unitization** have eliminated most of the negative effects of the Rule of Capture.

**Seaboard formula**—a rate structure for interstate gas pipeline tariffs that allocates 50 percent of the "fixed costs" to the "demand" component of customer charges and 50 percent to the "commodity" charge. COMPARE: **United formula; fixed-variable tariff.**

**scf**—standard cubic foot (sometimes called "cf" for short). A standard cubic foot of gas is the volume that would fill one cubic foot at 60 degrees Fahrenheit and at the atmospheric pressure found at sea level (14.73 pounds per square inch).

**secondary oil recovery**—techniques (primarily injection or reinjection of gas or water) for increasing the amount of crude oil that ultimately can be produced from a given pool.

**sedimentary basins**—large geographic areas that contain rock strata of a sedimentary nature, deposited when the topography was conducive to sedimentation, as in lakes or shallow seas.

**self-help program**—a federal program instituted during the gas shortages of the 1970s that offered industrial users the opportunity to utilize interstate gas pipelines as contract carriers when purchasing gas directly from producers in the field. As direct purchasers, self-help industrial customers were automatically exempted from wellhead price ceilings because the gas was not "for resale."

**service agreement**—the contract between a gas utility and a downstream purchaser that specifies all of the terms of service except the price itself, which is governed by a filed tariff that may from time to time be changed without the concurrence of the purchasing party.

**ship-or-pay**—a legal obligation borne by a gas company to make a specified payment to a gas pipeline or LNG tanker company in return for guaranteed transportation service. Like a "take-or-pay" condition imposed by producers on their pipeline purchasers, ship-or-pay obligations must be fulfilled whether or not the full service is actually utilized.

**shut-in gas**—a situation in which production is restrained either by order of a state conservation authority (pro-rationing) or because the producer is unable to find a buyer at an acceptable price.

**sinking fund**—a fund established in the interest of a bondholder to reduce the risks of loan default, wherein the borrower is required to deposit periodic repayments of debt principal into a special account that is transferred to the bondholder when the bond expires.

**slurry**—a mixture of solid products (like coal) with a liquid (such as water or oil) to facilitate movement through a fixed pipeline.

**source rock**—strata thought to have been the organic-rich parent rock for the genesis of a particular accumulation of gas and/or oil, which later migrated to a permeable "reservoir rock," capped by some sort of trapping mechanism that prevents further movement.

**spot market**—commodity transactions whereby participants make buy and sell commitments of relatively short duration, in contrast to the "contract" market in which transactions are long-term.

**substitute natural gas (SNG)**—energy-rich vapors manufactured from naphtha or propane in peakshaving plants to accommodate periods of greatest customer demand.

**supplemental gas**—year-round supplies of liquefied natural gas, high-btu coal gas, and Arctic gas sought by interstate pipelines during the supply shortages of the 1970s.

**surcharge**—a special customer charge that is not a standard part of a utility's tariff.

**swamp gas**—methane emanating from swampy or boggy areas. Swamp gas is produced by organic decomposition in an oxygen-poor environment.

**synthetic natural gas (SNG)**—energy-rich vapors manufactured from coal. Beginning in the 1970s, SNG projects were designed to produce high-btu grades of coal gas on a year-round basis.

**tariff**—a document filed by a gas pipeline or a gas distributor with its federal or state regulator, which sets forth the prices ("rate schedules") of its services to various classes of customers.

**take-or-pay**—a contractual obligation to pay for a certain threshold quantity of gas whether or not the buyer finds it possible (or beneficial) to take full delivery.

**therm**—100,000 btu.

**tectonic**—of or related to movement of the earth's crust, as in the faulting and folding attendant to mountain building.

**tight gas**—gas contained in rock strata with low permeability, thereby necessitating enhanced (and costly) production techniques like fracturing.

**tight sands**—sandstones rich in hydrocarbons but of low permeability.

**two-part tariff**—a tariff that includes both a "commodity" and a "demand" charge.

**unconventional gas**—gas whose costs of production (utilizing current technologies) would be higher than its current market value.

**United formula**—a rate structure for interstate gas pipeline tariffs that allocates only 25 percent of the "fixed costs" to the "demand" component of customer charges, and the remaining 75 percent to the "commodity" charge. COMPARE: **Seaboard formula; fixed-variable tariff.**

**unitization**—an agreement by all the multiple owners of a common reservoir to use joint production facilities and to apportion ownership of production and costs in a specified way. All the major oil-producing states except Texas and New Mexico authorize state oil and gas conservation agencies to impose a unitization plan if the owners do not arrive at a voluntary agreement.

**value-of-service pricing**—a method of apportioning costs among utility customers so that users who place a greater value on the service are charged higher rates than the more price-sensitive customers; sometimes called "Ramsey pricing."

**vanishing rate base**—the regulatory phenomenon by which a pipeline company's rate base (and hence its profits) will diminish through time (even if it suffers no decline in gas sales) unless it reinvests in new plant the capital it recovers through depreciation.

**variable costs**—those portions of a company's cost of service that will diminish if the service is curbed. For gas pipelines, variable costs include the cost of compressor fuel and (unless take-or-pay thresholds are in force) the cost of the gas commodity itself. COMPARE: **fixed costs.**

**vertical integration**—a situation in which a company and its affiliates are involved in more than one segment of a particular industry. In the gas industry, vertical integration exists if a company participates in two or more of the following activities: gas production, gas transmission, or gas distribution.

**vintage**—a regulatory category for natural gas based upon the year in which gas deliveries commenced for a particular well.

**warranty contracts**—gas contracts in which a producer commits to deliver a volume of gas with no specification as to the field of origin and with none of the production-related caveats that are commonly attached to contracts applicable to the sale of a particular gas reservoir.

**wildcat drilling**—drilling in a relatively unexplored geographic region or at an unexplored depth within a known hydrocarbon province. Wildcat drilling is riskier (resulting in more dry holes) than field delineation or field extension drilling.

**water gas**—an improved grade of coal gas produced in the early years of the gas industry by the injection of steam during the distillation process.

**zone tariff**—a gas pipeline tariff that provides service rates which vary between geographic units but are uniform within any single unit.

# INDEX

# ABOUT THE AUTHORS

Arlon R. Tussing and Connie C. Barlow operate an economic consulting practice, ARTA Inc., based in Seattle. Specializing in energy and natural resources, they have advised gas producers, utilities, and U.S. and Canadian federal, state, and provincial governments on natural gas issues. Recent topics have included the Alaska Highway gas pipeline, corporate acquisitions, utility rate design, and the development of a natural gas futures market. Tussing and Barlow are co-editors of ARTA's monthly publication, *Natural Gas Insights*.

Dr. Tussing is also professor of economics in the University of Alaska's Institute of Social and Economic Research. During the 1970s, he served as chief economist of the U.S. Senate Committee on Energy and Natural Resources. He has been a council member of the International Association of Energy Economists and has served on the advisory board of the U.S. Department of Commerce.

Before joining ARTA in 1978, Connie C. Barlow served in senior staff positions in Alaska state government, including the state senate and the Department of Natural Resources.